T0329488

EIGHTEEN HUNDRED AND
FROZE TO DEATH

EIGHTEEN HUNDRED AND FROZE TO DEATH

THE IMPACT OF AMERICA'S FIRST CLIMATE CRISIS

John V. H. Dippel

Algora Publishing
New York

Library of Congress Cataloging-in-Publication Data —

Dippel, John Van Houten, 1946-
    Eighteen hundred and froze to death: the impact of America's first climate crisis /
John V.H. Dippel.
        pages cm
    Includes bibliographical references and index.
    ISBN 978-1-62894-117-3 (soft cover: alkaline paper) — ISBN 978-1-62894-118-0
(hard cover: alkaline paper) — ISBN (invalid) 978-1-62894-119-7 (ebook) 1. Climatic
changes—Northeastern States—History—19th century. 2. Cold waves (Meteorology)—
Northeastern States—History—19th century. 3. Climatic changes—Social aspects—
Northeastern States—History—19th century. 4. Social change—Northeastern States—
History—19th century. 5. Northeastern States—Environmental conditions—History—
19th century. 6. Northeastern States—Social conditions—19th century. 7. Northeastern
States—Politics and government—19th century. 8. Northeastern States—Economic
conditions—19th century. I. Title. II. Title: 1800 and froze to death.
    QC903.2.U6D57 2015
    363.34'92—dc23
                                    2014049112

Cover art: "Farm in Winter," oil, by Emma May Seberry (1880–1918)

For Steve and Dave

Months that should be summer's prime
Sleet and snow and frost and rime
Air so cold you see your breath
Eighteen hundred and froze to death.

— old New England folk verse

Table of Contents

PREFACE

On the evening of April 10, 1815, fourteen-thousand-foot high Mount Tambora, situated on the Indonesian island of Sumbawa, blew apart in the largest volcanic eruption in recorded history — a colossal explosion ten times the size of the more famous one at Krakatoa decades later. Shock waves shook homes in the city of Batavia (present-day Jakarta), eight hundred miles away. Some twenty-four cubic miles of molten magma and pulverized rock were violently spewed into the atmosphere. A dark column of ash and smoke shot up some thirty miles and quickly spread across the Pacific sky, for two days plunging the ocean and islands below into total darkness over a radius of 370 miles. Floating islands of pumice several miles long and over three feet thick covered the surrounding sea. Gas emitted in the explosion reacted with atmospheric water vapor to form droplets of sulfuric acid. These circled the globe, reflecting back sunlight and causing surface temperatures to fall.[1] An estimated seventy-one thousand persons living near Tambora died as a result of this explosion — due to superheated gases, powerful whirlwinds, tsunami waves, starvation, and disease.[2]

But this was only the beginning of this eruption's impact on life and the

---

[1] The prominent role played by volcanic eruptions in lowering surface temperatures on earth has been recently confirmed by scientists involved in the Berkeley Earth Land Temperature Project. See, for example, Robert Rohde, et al., "A New Estimate of the Earth Surface Land Temperature Spanning 1753 to 2011," *Journal of Geophysical Research* (submitted July 2012). http://berkeleyearth.org/pdf/results-paper-july-8.pdf This study concludes that increased level of "volcanism" is mainly responsible for the "sudden drops" in temperature during the years 1753-1850.

[2] For these details, see Jelle Zeilinga de Boer and Donald T. Sanders, *Volcanoes in Human History: The Far-Reaching Effects of Major Eruptions* (Princeton: Princeton University Press, 2002), 139-47.

course of human history. Back then, no one in North America or northern Europe had any idea what had happened on this remote island. A year later, however, millions in those parts of the world would find themselves confronted with its dire consequences. This is the story of what they faced, how they responded to this unprecedented natural crisis, known in New England as "Eighteen Hundred and Froze to Death," and, more generally, as "the year without summer, "and how the ensuing changes in outlook helped to speed up the process of modernization in the United States.

Introduction

"It is the common exclamation, what is to become of us?"
—*letter in the Boston Gazette, 15 July 1816*

He pulled the door open and stepped outside. At once he was struck by a blast of cold, wintry air, which stung his face and bare hands like needles. He stopped and stared. Two inches of fresh snow blanketed his yard. Snow stretched across the fields and ran up the slopes of the Green Mountains to the west. Heavy snow draped the peaks. Shaking his head, he turned and trudged slowly across the frozen ground to his vegetable garden. The short cornstalks were bent over by snow and ice. Underneath they were shriveled and blackened. The dangling peapods were just as bad. When he dusted them off, they looked as if they had been scorched by fire.

He glanced up at the peach trees, but the blossoms had frozen solid. There was no saving them either. All his crops were ruined. If the ground thawed, he could start to plant all over again, but there was too little time for anything to ripen before the first frost of the fall.

It was only the beginning of June, but it looked as if summer had already come and gone. In his thirty years in Vermont, this was the strangest, most forbidding year he had ever seen. And the worst was yet to come.

Indeed, 1816 would turn out to be the most ruinous year in the recorded history of New England. During that startlingly cold summer, repeated hard frosts, subfreezing temperatures, and snowfall would kill off much of the region's harvest — upon which both life and livelihood depended. This untimely destruction of vegetables, grains, and fruits would make it impossible for many subsistence farmers, especially in northern New England, to feed their families.

Many would be threatened with starvation and be forced to pack up their few belongings, abandon their lands, and move west. This would produce one of the largest human migrations that America has ever witnessed — a steady stream of wagons, oxcarts, horses, and distraught, downcast Yankees trudging across New York and Pennsylvania to the promised land of Ohio. Entire villages lost much of their population, and some remained ghost towns for decades. Those families which stayed behind could only survive by eating half rotten corn, soggy potatoes, and boiled nettles. Their sheep froze to death, and their cattle perished the following winter because there wasn't any hay to eat. It was so cold in some parts that migratory birds froze and dropped dead in the streets like stones.

Along with these severe physical hardships, the people of New England faced a spiritual dilemma. Descendants of Puritans who had come to the New World believing this was the mission God had chosen for them, they now had to wonder what had provoked this awful display of divine displeasure. Why, for the first time in living memory, had the Deity failed to keep his promise to provide a bountiful harvest? What was their sin, and how could they atone for it? How could such a catastrophe befall a people who had bent to His will so dutifully? Had God abruptly abandoned them? The bizarre, unprecedented weather of 1816 even caused some to question the central Christian belief that the Almighty had created the earth and the heavens in His image, and that all events unfolded in accordance with His benevolent guidance. Was it possible that God did not control the wind, the rain, the storms, the snowfall, volcanoes, earthquakes, and the change of seasons — let alone each and every sparrow which plummeted to the earth?

In this regard, the cold summer of 1816 was as unsettling — and controversial — as the more prolonged "climate change" we are now experiencing. Within a relatively short period of time, dramatic, unmistakable shifts in the weather occurred, creating havoc, destroying crops, forcing an exodus, and endangering the lives of many thousands, if not millions, of persons. (As will be shown in this book, the cold spells of 1816 were not an isolated aberration, but part of a long-term cooling trend. Furthermore, New England was not alone in experiencing unusually low temperatures that summer: much of northern Europe and places as far away as China were also affected. Food shortages led to riots, malnutrition and starvation, higher crime rates, the spreading of diseases like typhus, and expanded social welfare programs.) Then, as now, this change provoked much heated debate, since this prolonged chill affected so many people, calling into question their core assumptions about the nature of our world and the forces which govern it. Inadequate knowledge about the climate's creation invited all manner of speculation, while stirring fears about what was yet to come. At the same time, heightened interest in understanding and predicting the weather encouraged new ways of studying it.

Inevitably, such closer scrutiny yielded insights which warred with longstanding conventional wisdom. Specifically, scientific investigation of the weather heightened latent tensions between faith and reason. Because of their severity and because they seemed to defy biblical teachings, the unseasonal killing frosts of 1816 induced some to seek physical — rather than metaphysical — explanations. The scope of these crop failures — all the way from northern Maine to Alabama — seemed to divorce them from any focused, punitive purpose. (Could God be so indiscriminately wrathful and still remain just?) Regular, accurate measuring of temperatures, barometric pressure, snowfall, and ice thickness tended to objectify these phenomena and detach them from the sphere of morality. Consequently, they assumed the guise of merely *natural* happenings, not the visible manifestation of godly retribution. Reinforcing this perception, the frosts and drought that summer proved impervious to human supplication: the prayers offered up by the faithful gathered in New England's whitewashed churches failed to end them. This failure was deeply troubling because it suggested that the deity was not listening. Indeed, it made some wonder if God existed at all, or, if He did, how much control over earthly events He really had. If God's hand did not stir the winds, unleash the rains, or cause them to cease — if He did not convey his wrath through earthquakes, floods, and volcanic eruptions — then *Homo sapiens* was utterly alone, at the mercy of these disasters. To acknowledge that they occurred randomly and unintentionally was to call into question the very meaning of human existence. But other New Englanders responded to this crisis by redoubling their efforts to atone for it. They took the calamitous weather as a sign of grievous error on their part — another deserved fall from grace. It shook them out of their spiritual torpor, reminding them of the grave consequences of neglecting the Sabbath and disobeying God.

Overall, the apparent disconnect between natural events and super-natural purpose weakened the case for faith. For how could the harsh, destructive weather of 1816 be squared with divine benevolence? Thus, some New Englanders — particularly the better-educated ones — became more amenable to alternative, rational explanations. Building upon the experiments of men like Benjamin Franklin (who had demonstrated that lightning rods did a better job of protecting churches than prayer), they began to explore the world around them with unfettered minds. They hoped to find answers through scientific description and analysis, instead of relying on the word of the clergy. Their meticulously recorded observations would constitute the raw data for wholly new, revolutionary theories not only about the weather but about all earthly processes and events. As this inquiring spirit took hold, the wisdom of Puritan forefathers increasingly seemed inadequate and outmoded. More generally, trust in higher authority was weakened. What had begun as a climate crisis quickly evolved into a crisis in belief.

This development had broad consequences. New England had long been a bastion of unshakable faith in established ways and infallible authorities. Connecticut — perhaps the most hidebound state in the region — was proud to call itself the "land of steady habits." Here, as elsewhere in early nineteenth-century New England, a Congregationalist theocracy reigned, its monopoly on ecclesiastical power and knowledge of divine intent virtually unchallenged. But the inability of its ministers to explain why God was so wantonly and indiscriminately punishing so many devout believers and to obtain relief through acts of contrition undermined their stature. This evidence of their helplessness cleared the way for other denominations — Methodists, Baptists, Unitarians, Quakers, and other so-called "infidels" — to establish footholds across New England. It also gave a boost to uneducated, backwoods preachers who now dared to deny that only one clerical elite could explain the ways of God to man. As a result, religion in the young American republic was thoroughly "democratized" — rendered more open, pluralistic, and egalitarian. "Tolerance" took the place of sectarian deference.

A similar transition took place in the political arena. Federalist power — all but eliminated in other parts of the country — had long been intimately connected to the Congregationalist "standing order," with ministers like Yale's Timothy Dwight serving simultaneously as the "pope" of his church in Connecticut and de facto head of the state's ruling party. Disenchantment with the clerical establishment after 1816 was paralleled by the demise of Federalist hegemony in several New England states. Radical notions about taxation, suffrage, judicial independence, and democratic governance toppled the old hierarchical order. The same transformation took place in the region's eminent colleges and universities: instead of only preparing young men for careers as Congregationalist ministers, these institutions embraced a more secular academic mission, adding courses in mathematics, astronomy, and physiology to their curricula and admitting students of other faiths to their classrooms.

More dramatic than these structural changes in New England was the human flood westward. The crop failures of 1816 forced tens of thousands of families to leave the rockbound fields of their forebears and seek better fortunes inland. This westward shift in population diminished the region's economic and political clout and permanently altered the national balance of power. As the territories and new states of the Ohio Valley grew and prospered, the countryside and towns of New England were vacated and depleted.

It would be wrong to construe the catastrophic weather of 1816 as the *cause* of all these changes. It happened to occur at a propitious moment in American history, when new values and way of life were taking shape after a period of post-colonial stagnation. More than two decades after the Revolution, the young republic was still largely steeped in the traditions of its former colonial master. Especially in New England, political, religious, educational, and social hierarchies imported from Britain still held sway. Methods of farming had not changed much from the days when the Pilgrims broke the rocky soil around Plymouth and planted their first kernels of corn in the 1620s.[1]

But longing for a genuinely American ethos was starting to emerge: a desire that the United States live up to the lofty promise of the Declaration of Independence and institute popular forms of governance; a yearning to explore the continent and make a fresh start on the frontier; an eagerness to increase one's standard of living through new, money-making endeavors; a curiosity about the natural world and how to exploit its resources; and a confidence in one's own opinions and experience in lieu of trust in one's "betters." The "year without a summer" was a catalyst — an accelerant — for this already developing process of national transformation. If it had not occurred, the nation would still have made this transition into a more rational, democratic, modern era, but at a slower pace.

In New England, the year 1816 is remembered mainly for its harmful, puzzling weather. But, in an overwhelmingly agricultural country such as the United States was in the first part of the nineteenth century, a sharp deviation in the weather pattern was more than a curiosity or inconvenience. Survival utterly depended upon a predictable, reliable cycle of seasons and the harvests this brought. When this did not take place, not only was an entire way of life put in jeopardy, but also a whole raft of beliefs — in the future, in the benign aspect of Nature, in the omnipotence of a divine creator, and in the earthly authorities who professed to understand His ways — were shaken.

This book analyzes the impact of the "year without a summer" on several aspects of early nineteenth-century American life — Western settlement, democratization, religious belief, scientific inquiry, industrialization, and philosophical outlook. Each chapter examines the status quo in these spheres just prior to 1816 and suggests how these situations and outlooks were affected by the disastrous cold spells experienced that spring and summer.

---

[1] A major change occurred in 1623, when these farmers gave up communal farming and planted individual gardens. This switch — instigated by Governor William Bradford — saved the colonists from starvation.

The consequences of this extraordinary year would reverberate for decades to come — indeed, to our own day. For the lessons of 1816 are ones we are still attempting to absorb — namely, the implications of living in an uncertain, volatile universe, coping with unforeseen, life-threatening change, and accepting the limits of human capacity to control natural forces. As we wrestle to find the appropriate ways to respond to climate change in our own day, we can benefit from looking closely at how our ancestors responded to a similar challenge. While it might be tempting to conclude that we have much more working for us than they did — reams of scientific data, highly sophisticated computer modeling, a firm grasp of weather patterns and disturbances, a knowledge of the earth's climatic history stretching back over eons — it would be foolish for us to think that the same conflicts between reason and belief, between fear and hope, between the need for certainty and limited understanding, do not shape our reaction much as it did theirs. In this sense, we are still living in the world of 1816, and it still has much to teach us.

Chapter 1. The Coldest Summer on Record

While the earth remains, seedtime and harvest,
cold and heat, summer and winter,
day and night, shall not cease.

— Genesis 8:22

The surprisingly inhospitable weather of 1816 was deeply unsettling to New Englanders, as it weakened their trust in a benign, protective Providence and forced many to contemplate a radical change in their lives.

After a relatively mild winter, spring arrived that year in its usual disguise — as a seductive flirt. If anything, it was a little ahead of schedule. In the Berkshires, robins were spotted as early as the end of February, although, when the temperature dropped, they disappeared, not to be seen again for weeks.[1] The blustery and raw days of March, which a querulous Amanda Elliott, twenty eight years old and still living at home with her parents on a spit of land jutting out into the Long Island Sound, had pronounced "cold as Greenland," finally loosened their icy grip.[2] In early April, the days grew warmer. Ice melted away. Warm rain stirred green shoots to life. People passing in the streets nodded at each other. The long winter of Puritan forbearance was coming to an end. The hardy, stoic people of New England had survived another wintry test. Inwardly, they rejoiced, knowing it might have been much worse. Jonathan Edwards, dead

---

[1] David D. Field and Chester Dewey, *A History of the County of Berkshire, Massachusetts, in Two Parts* (Pittsfield MA: S.W. Bush, 1829), 183.
[2] Amanda Elliott, "Diary of Amanda Elliott," entry for 18 March 1816, typescript (photocopy), 920 EL 425, History and Genealogy Unit, Connecticut State Library, Hartford.

now over half a century, had taught them that God could — if He so chose — punish even the most devout believers. Still, such afflictions were nothing compared to what would befall the truly wicked, when the Lord unleashed His full wrath upon them, melting their bodies — in Edwards' vivid simile — like silver in a fiery furnace.[1] Now, as the trees budded, the charitable aspect of the Almighty was revealing itself — providing the bounties of the earth, the fruits and grains and vegetables which would sustain them through the summer and the long, cold, dark months which would follow.

Then, like a shy lover, the spring faltered and retreated. Stern, harsh winter reasserted its prerogatives. Overnight temperatures plummeted. Frost shocked the new growth, arresting its progress and rotting it. In central Connecticut, the Reverend Thomas Robbins was too preoccupied by another outbreak of typhus, as well as severe back pain, to dwell upon this odd weather pattern.[2] Nearly every day he rode out in his sleigh to visit sick and dying parishioners — mostly women, the elderly, and children — and gave them words of comfort. Over the wailing, feverish infants, writhing in their cribs, he could only silently pray. Each week his East Windsor congregation was dwindling — five members had been lost between two communions held at the meeting house during March. Victims of God's apparent displeasure, many of them remained mired in "stupidity" to the end, unable to welcome the Lord into their hearts. Was there some connection between their stubborn neglect of the Sabbath, God's withholding of spring, and the procession of coffined corpses being borne into his church, one or two at a time? The Reverend Robbins could not answer this question. To assuage his nerves, he immersed himself in routine tasks. For he had his own practical needs to address. God would not provide to one who did not sow his own seed — even a man of the cloth. Growing one's own food was central to the Puritan ethos of austere self-reliance. As one Massachusetts manufacturer would later recall, "A man who did not have a large garden of potatoes, crook-neck squashes, and other vegetables, besides a hog in the sty — fattened to fill his pork barrel on each Thanksgiving — was regarded improvident."[3] So he took advantage of a brief warm spell in early April to plant peas in his garden, as he always did at this time of year. But, around the eleventh, heavy snow fell, not only in many parts of New England, but

---

[1] "Of the Annihilation of the Wicked," *Rhode-Island American*, 16 March 1816. This article cites Edwards's sermon, "The Eternity of Hell Torments," of April 1739.

[2] Several outbreaks of typhus, associated with cold, damp weather as well as poor hygiene, would be reported in 1816. See, for example, "Climate of the United States," reprinted from *Niles Weekly Register* in the *New-Jersey Journal* (Elizabethtown NJ), 20 August 1816.

[3] Edmund S. Hunt, *Weymouth Ways and Weymouth People: Reminiscences* (Boston: privately printed, 1907), 16.

across western New York and as far south as Washington, D.C. When the Reverend Robbins swept dead leaves off his asparagus beds a few days after this storm had ended, he found that not a single spear had come up. But this did not trouble him much. He knew that God would still provide. His children only had to remain patient. And, indeed, by the end of the month, the sun had returned, and the pungent earth, recently hard as a rock, yielded up its promised reward. For supper he devoured his first asparagus of the season.

Still, up in Cambridge, when Harvard professor of mathematics John Farrar tabulated his thermometer readings for the entire month, he found that this April had been two-and-a-half degrees colder than usual. This was surprising, but no cause for concern. Farrar knew that the New England spring was invariably fickle. For days, it might drizzle, the air raw and near freezing under impenetrable gray skies, and then the sun would suddenly burst forth and seduce the earth into fecundity, only to go away again just as quickly, leaving the exposed plants and flowers to stiffen in the cold breeze. In recent years, such "wintry springs" had become the norm in this part of the world.[1]

The plentiful rain which fell early in May seemed a harbinger of God's promise finally being fulfilled. Writing from Litchfield, in the mud-slick northwest corner of Connecticut, a relieved Benjamin Tallmadge could boast that he and his neighbors had "no needs," as "blessed" showers had thoroughly soaked their fields. These "genial visitations" augured well for spring planting.[2] However, all was not so rosy further to the north, where winter tarried like a gruff elderly relative — preventing everyone else from carrying on with their lives. Once again, new buds were entombed in sheaths of ice. This next cold spell spread down to the Deep South, killing off cotton in faraway Tennessee.

May was marred by more trying reversals. Strong winds roared southward over the St. Lawrence River bringing another destructive cold snap. By the middle of the month, ice encased fruit trees as far south as Pennsylvania and Virginia.[3] Most of the Indian corn that had been planted earlier was destroyed. This was a serious blow, as it was the mainstay of the New England diet, mashed into cereal or ground into flour to make hearty dark bread. So were garden vegetables. In East Windsor, their blossoms were so decimated by hard frosts that the Reverend Robbins feared for their survival.[4] The cold and

---

[1] "The Season," *Providence Patriot*, 8 June 1816.
[2] Letter of Benjamin Tallmadge to John P. Cushman, 2 May 1816, Benjamin Tallmadge Collection, Series 1, Folder 36, Ingraham Memorial Library, Litchfield Historical Society, Litchfield CT.
[3] Keith C. Heidorn, "Eighteen Hundred and Froze to Death: The Year There Was No Summer." http://www.islandnet.com/~see/weather/history/1816.htm.
[4] Thomas Robbins, *Diary of Thomas Robbins, D.D., vol. 1, 1796-1854* (Boston:

drought persisted for most of the month. On the fifteenth, for example, it was so frosty in the southern New Hampshire village of Old Chester that a grown man could walk across the ice covering his fields without breaking it.[1] Then, only three weeks later, the temperature there soared to eighty-eight degrees. But elsewhere in the country, the weather took a turn for the worse. At the beginning of May peach trees in the Ohio outpost of Chillicothe, which had flowered at the end of March, were dusted with snow; three inches had piled up before it was all over. [2] Back east, there were more signs that this season was behaving strangely. After seeming to have finally spent itself, winter returned with a vengeance in the last days of May. Unseasonably chilly weather was experienced from snowy Quebec to Erie, Pennsylvania. In the wilderness west of the Adirondacks, where Noadiah Hubbard had been trying to start a new life as a land agent and storekeeper, he complained that the weather was still "cold and backward" — an alarming development after the poor wheat season the year before.[3] When Harvard's John Farrar tallied up the numbers, he calculated this was the third coldest May on campus in a quarter century.[4]

Meanwhile, Yankee farmers bided their time, waiting patiently to do another spring planting. They knew full well that, in New England, "winter lingers in the lap of spring."[5] The break finally came the first days of June, when temperatures rose into the high seventies and eighties all the way from Williamstown, Massachusetts, to New Haven, Connecticut. [6] In Salem, it was even hotter — more than one hundred degrees for three days in a row.[7] Stripping off their woolen overcoats, stern-faced, grizzled men strode out to their moist fields under a steady sun to break up the soil with wood plows and then mold it into small mounds, inserting three or four dried kernels in the middle, as the Patuxet and Narragansett tribes had taught their Puritan

---

Beacon, 1884), 661-70.

[1] Benjamin Chase, *History of Old Chester (N.H.) from 1719 to 1869* (Auburn NH: Benjamin Chase, 1869), 173.

[2] "The Weather," *Providence Gazette*, 11 May 1816. The report from Ohio was dated 11 April.

[3] Letter of Noadiah Hubbard to Julius Deming, 28 May 1816. "Hubbard, Noadiah" folder, Quincy Family Papers, Litchfield Historical Society. For more details on Hubbard's business activities, see John A. Haddock, *The Growth of a Century: As Illustrated in the History of Jefferson County, New York, from 1793 to 1894* (Albany NY: Weed-Parsons, 1895), 511.

[4] Don Sutherland, "The June 1816 Snowfalls and the Cold Summer of 1816," 2-3. http://www.theedwardbulwer-lyttoninstitute.org/uploads/1/7/5/8/1758444/summer1816.pdf.

[5] Sidney Perley, *Historic Storms of New England* (Salem MA: Salem Press, 1891), 206.

[6] The temperature in New Haven reached 79 degrees on June fifth. Diana Ross McCain, "Year Without Summer," *Connecticut* (July 1987): 49.

[7] Perley, *Historic Storms*, 206.

ancestors to do. But this warm spell did not last either. On the fourth of June, another blast of cold Canadian air caused thermometers to plunge. Sultry Salem had a dusting of snow on the eighth, and Williamstown was covered with a foot.[1] In Cabot, in northern Vermont, the snow was a foot and a half deep.[2] Along the usually temperate Long Island Sound, the mercury fell forty five degrees in less than a day and a half. Amanda Elliott, who had been overcome by heat during a church service a few days before, was so unnerved by this surprising turn of events that she stopped writing in her diary for three weeks. The massive front rapidly spread over the rest of New England, dumping heavy snow and making a mockery of the Farmer's Almanac prediction that June would be "very warm."[3] By the sixth, this snow cover extended across most of northern Maine, New Hampshire, and Vermont, and into the northwest corner of Massachusetts.[4] In Montpelier, it was up to a man's knees.[5] On the eighth, another six inches fell on parts of Vermont, and now the whole region was as frozen and forlorn as in the depths of winter.

Plants shriveled and died — most alarmingly, the young corn. Leaves on some trees stiffened and died. The beech trees stood bare and gaunt as they did in January. And, still, the snow kept coming. It fell for four days without stopping in Vermont and in the neighboring parts of Maine and New Hampshire. It fell on the Catskills, on the sawmills at Bangor, on Benjamin Harwood's farm in Bennington, outside the Plymouth, Connecticut, shop where a young clockmaker named Chauncey Jerome was hard at work, and on the roof of the imposing Georgian house on Essex Street in Salem, Massachusetts, where the Reverend William Bentley had been a boarder for the past quarter century. A howling wind whipped it around. It blew so hard that guests walking to the meetinghouse in Concord, New Hampshire, to witness the swearing-in ceremony for the newly re-elected governor,

---

[1] Ibid., 208.

[2] "The Season," *Reporter* (Brattleboro VT), 17 July 1816.

[3] McCain, "Year Without a Summer," 48. There are stories about a mischievous copy editor inserting the prediction that it would snow on July 13, 1816, into that year's almanac the preceding fall, when the founding editor, Robert B. Thomas, was sick in bed. According to these tales, an irate Thomas had almost all of the copies destroyed, but some survived, with the ironical accurate forecast preserved. But none of these "original" versions of the 1816 almanac has ever come to light. Many other almanacs were published back then, but their predictions for June only conformed to what was to be expected at that time of year.

[4] For a map of this snow cover, see Henry Stommel and Elizabeth Stommel, *Volcano Weather: The Story of 1816, the Year Without a Summer* (Newport RI: Seven Seas Press, 1983), 28-9.

[5] Barrows Mussey, "Yankee Chills, Ohio Fever," *New England Quarterly* 22:4 (December 1949): 435.

William Plumer, could barely make any headway against it.[1] Snow squalls in Sanbornton, New Hampshire, were so thick and relentless that workmen cutting down trees for a schoolhouse had to call it quits and retreat indoors to warm their hands. In the White Mountain hamlet of Warren, further to the north, the "whole face of Nature appeared shrouded in gloom. The lamps of Heaven kept their orbits, but their light was cheerless. The bosom of an earth in Midsummer's day was covered by a wintry mantle, and man, and beast, and bird, sickened at the prospect . . . it seemed as if the order of the season was being reversed."[2] An article in an upstate New York newspaper confirmed what many longtime residents were saying: no one had ever witnessed a spring like this.[3] During a stopover in Pittsburgh on his way to Ohio, the farmer David Thomas encountered "brisk gales" from the northwest and the same blanketing frost he had already run into on the road west for several days in a row.[4] Connecticut's last Federalist governor, John Cotton Smith, recorded that the ice in his home town of Sharon, not far from the Massachusetts border, was almost as thick as his little finger.[5] Men laying bricks in Bath, New Hampshire, beside the roaring Ammonoosuc River, had to give up because their tub of mortar kept freezing. On June 10th, Chauncey Jerome, trudging to work in Plymouth, had to stop, lay down his tools, and put on mittens in order to keep his fingers from going numb.[6] Outside the town of Peacham, Vermont, on the far side of the White Mountains, an eighty-four-year-old man named Joseph Walker got lost in the unexpected snowfall and ended up spending a night alone in the woods. When he was found the next day, Walker was so frost-bitten that that one of his toes had to be cut off.[7] Up in Hallowell, Maine, near Augusta, migrating hummingbirds, finches, Purple Martins, Scarlet Tanagers were so "benumbed" by the unfamiliar cold they could be held in a cupped hand like

---

[1] Stommel and Stommel, *Volcano Weather*, 30. The memoir from which this anecdote is drawn may have been slightly exaggerated.

[2] William Little, *A History of the Town of Warren, New Hampshire, From its Early Settlement to the Year 1854, including a Sketch of the Pemigewasset Indians* (Concord NH: McFarland and Jenks, 1854), 90.

[3] Untitled article, datelined Watertown, New York, *Albany Advertiser*, 7 June 1816.

[4] David Thomas, *Travels through the Western Country in the Summer of 1816* (Auburn NY: David Rumsey, 1819), 56.

[5] McCain, "Year Without a Summer," 50.

[6] Chauncey Jerome, *History of the American Clock Business for the Past Sixty Years, and a Life of Chauncey Jerome, Written by Himself* (New Haven: Drayton, 1860), 31.

[7] A similar mishap befell a Maine farmer out looking for his sheep. He was found three days later, alive but with his legs frozen. See passage from anonymous Fryeburg, Maine, diary. Quoted in Lee-Lee Schlegel, "The Year Without a Summer: 1816, in Maine." http://www.milbridgehistoricalsociety.org/previous/no_summer.html.

newborn chicks — before they died.[1] Recently shorn Merino sheep, recently brought over to the New World from Spain, froze to death standing in their pens, despite efforts to cover them with their own fleece.[2] As one Vermont journalist summed up, this was "indeed a gloomy and tedious period."[3]

Huddled indoors, farmers could only stoke their fires, suck on their corncob pipes, mutter, and shake their heads.[4] Much, if not all, of their newly planted corn was "cut off." Normally, they would store enough in chests to feed a family for the half a year when it was too cold for crops to grow.[5] But the June cold snaps had made it unlikely these reserves would suffice.[6] On top of this, squash, beans, and other vegetables had also been killed off by the heavy frost and ice. In wheat- and rye-growing regions like New York's Hudson Valley, those cereal grains suffered extensive damage.[7] Families in northern settlements, where the growing season was shorter, had little hope of any harvest at all. This meant they would not have much to eat. In 1816, there were few general stores in rural New England, no wagons arriving overland from warmer climes bearing bags of rice or barrels full of flour, no railroad cars or canal boats to bring in tons of supplies to make up for what they had lost. People who lived by the coast could count on fish to help sustain them, but for most New Englanders this was not an option.[8] Dairies in Berkshire County, Massachusetts, and around Goshen, Connecticut, produced plenty of cheese — some half a million pounds in 1810 — as well as butter.[9] Well-off families had caches of these commodities. Some others had stashes of salt pork and beef in the cellar to help them through the long winter.[10] But none of these foodstuffs could avert what looked like a looming famine. Uncertainty

---

[1] "Hallowell, June 12," *Connecticut Gazette* (New London CT), 26 June 1816.

[2] Mussey, "Yankee Chills," 435.

[3] "Melancholy Weather," *North Star* (Danville VT), 15 June 1816.

[4] Isaiah Thomas, *The Diary of Isaiah Thomas, 1805-1818*, ed. Benjamin Thomas Hill (Worcester MA: The Society, 1909), entries for 12 and 18 June 1816, 316. Cf. Robbins, *Diary*, entry for 7 June 1816, 670.

[5] Sarah McMahon, "A Comfortable Subsistence: The Changing Composition of Diet in Rural New England, 1620-1840," *William and Mary Quarterly* 42:1 (January 1985): 31-2.

[6] McMahon, "'All Things in Their Proper Season': Seasonal Rhythms of Diet in Nineteenth Century New England," *Agricultural History* 63:2 (Spring 1989): 132, 135.

[7] "From the *Dutchess Observer*," *National Advocate*, 14 June 1816. Quoted in Sutherland, "June 1816 Snowfalls," 8.

[8] In coastal New Hampshire, farmers unable to fatten up pigs for slaughter due to a corn shortage resorted to catching mackerel instead. Perley, *Historic Storms*, 204.

[9] Howard S. Russell, *A Long, Deep Furrow: Three Centuries of Farming in New England*, abr. ed. (Hanover NH: University Press of New England, 1882), 145. For many decades "cheese wagons" from Goshen were a common sight in surrounding Connecticut villages; for half a century, the town was known as the nation's "cheese capital."

[10] McMahon, "Comfortable Subsistence," 45, 48.

and fear were deepening. "What is to become of this country, it is impossible to divine," noted a correspondent for the Boston-based *Centinel*, writing from Jackson, Maine, where the blossoms on some five hundred newly planted apple trees had just been destroyed. Sobered by this sight, he described the situation as "distressing beyond description" and predicted a massive food shortage throughout this part of Massachusetts and beyond. [1]

Judging by such newspaper reports, this was a massive aberration in the weather, extending over most of the country and abroad. Many speculated about why this was happening. Using newly refined instruments like the thermometer and the telescope, a handful of "natural philosophers" on both sides of the Atlantic were now able to collect accurate meteorological data and closely follow the movement of stars and planets; these observations led them to propose causal connections between terrestrial and heavenly occurrences. Some noted there had been a total eclipse of the moon on June ninth, as well as an eclipse of the sun two weeks earlier, suggesting the confluence of these celestial events might have somehow conspired to stop the advance of spring, perhaps by altering the direction of the winds. Others pointed out that sunspots, clearly visible to the naked eye, had been observed since the end of April. [2] An article in a London newspaper described six spots as looking like a "cluster of islands." They were estimated to be at least as large as the earth in diameter. These sunspots had appeared shortly after abrupt drops in temperature. [3] It was suggested that these black dots were preventing some of the sun's rays from reaching the fields of New England. Another theory held that sunspots had "extracted" heat from the earth, thus making the air temperature "less warm and genial." [4] Some argued that the earth had been slowly cooling over the past millennium, as its core temperature declined. The greater-than-usual number of ice floes spotted in the North Atlantic lent credence to this theory. [5] (It was believed that these ice masses absorbed the sun's heat and further lowered temperatures on the land. [6]) Proponents of this theory noted that this decline in temperature

---

[1] Untitled article from *Boston Centinel*, 22 June 1816, reprinted in *Connecticut Courant* (Hartford), 25 June 1816. Maine did not become a state until 1820.
[2] Thomas, *Diary*, 311. Cf. Joseph B. Hoyt, "The Cold Summer of 1816," *Annals of the Association of American Geographers* 48:2 (June 1958): 121. In Plymouth, Massachusetts, Leonard Hill recorded he had spotted five separate groupings of sunspots between April and August. Hill, *Meteorological and Chronological Register* (Plymouth, MA: Moses Bates, 1869), 62-4.
[3] "Domestic Summary," *Vermont Gazette* (Bennington VT), 2 July 1816.
[4] Untitled article, *Ulster Plebian* (Kingston NY), 25 June 1816. See also "Solar Macula," *National Daily Intelligencer* (Washington DC), 24 July 1816.
[5] Stommel and Stommel, *Volcano Weather*, 43. Cf. "The Season," *Reporter*, 17 July 1816.
[6] C. Edward Skeen , *1816: America Rising* (Lexington: University Press of Kentucky, 2003), 32. Conversely, some believed, large areas of forest, as then existed in the United States, tended to absorb the sun's heat and increase

had accelerated in recent years: the unseasonably cold weather of 1816 was its most dramatic manifestation.[1] Other scientific thinkers believed that earthquakes restored an equal diffusion of "electrical fluid" in the atmosphere, which, in turn, produced lower than normal temperatures. The fact that major earthquakes had occurred in the three preceding years seemed to back up this contention.[2]

These ruminations of the educated elite were based on little knowledge and only a nodding acquaintance with the writings of contemporary astronomers like William Herschel. (In 1801, Herschel, a German immigrant who had been one of the first to observe Uranus, had published a paper postulating that sunspots indicated an *increase*, not a decrease, in solar radiation.[3]) Most of their friends and associates scoffed at these explanations, as did the general public.[4] Even those with some understanding of the natural world were dubious. A future horticulturist, David Thomas pooh-poohed the idea that extraterrestrial events like an eclipse could affect the weather on earth, calling this an "absurdity."[5]

For most New Englanders, scientific investigation into meteorological phenomena was a waste of time. It was self-evident to them that God alone controlled the weather. In His infinite and benevolent wisdom, He had created a world where beautiful flowers bloomed each spring, fruit ripened luxuriously on heavily laden trees, and billowing, tasseled fields offered up what man needed to subsist, year after year, without fail. So it was written in Genesis, and so it had always been. They could count on a protective deity sustaining the cycle of the seasons much as they took for granted the sun's rising in the east each morning. True, God might occasionally punish some groups or individuals for sinful behavior — by causing storms, earthquakes, volcanic eruptions, drought, and other destructive events. (For example,

---

temperatures. See, for example, Timothy Dwight, *Travels in New-England and New-York*, vol. 1 (New Haven: Timothy Dwight, 1821), 60-61. Dwight did not wholly subscribe to this theory, however. Nor did he believe that the New England temperatures were becoming milder.

[1] John D. Post, *The Last Great Subsistence Crisis in the Western World* (Baltimore: Johns Hopkins University Press, 1977), 14. The years 1812-1818 were marked by lower average temperatures than the preceding and following periods. Cf. Chester Dewey, "Result of Meteorological Observations, Made at Williams College," *North American Review* 5:30 (21 May 1817): 7.

[2] For an elaboration of these theories, see "On the Cold of the Present Season," *National Daily Intelligencer*, 30 September 1816.

[3] Herschel pointed to an inverse relationship between their prevalence and the price of wheat in England. For a discussion of his sunspot theory and the contemporary evidence apparently refuting it, see "Spots on the Sun," *North American Review* 3:8 (July 1816): 285-7.

[4] See, for example, the mockingly dismissive comments about sunspot theories in "Political Humor," *Providence Gazette*, 22 June 1816.

[5] Thomas, *Travels*, 56.

some New England Federalists believed God was now expressing His anger at America for having gone to war against England.[1]) These were signs of divine unhappiness, which required acts of contrition. Still, the Puritans believed they were God's chosen people, whom He would safeguard despite their occasional failings.[2] That the fecund spring warmth was now being withheld — for no apparent reason — was thus profoundly unsettling. Had God broken His contract? Did the pious people of New England no longer enjoy His favor?

It is impossible to gauge how many people in New England began to question their faith because of these crop failures and continuing cold. Few jotted down their doubts in letters to friends or relatives, for to do so was to invite condemnation for blasphemy. Few dared express them to their neighbors, certainly not to their minister. One can only infer that there was some growing disbelief from the number of newspaper commentaries in affirming that God had *not* forsaken the faithful. As early as June twelfth, the editor of the *Hampshire Gazette*, in Northampton, Massachusetts, reassured his readers that the "untimely" weather of late, with its killing frost and deep snows, was no cause for alarm: "there will yet be a great plenty," the article affirmed, paraphrasing from the first letter of Paul to the Corinthians: "Paul may plant, and Apollo may water, but God giveth the increase."[3] After enduring eight nights of hard frost in early June and seeing their "tender vegetable tribe" destroyed, some Massachusetts residents had apparently begun to question why God was punishing them because Benjamin Russell, editor of the staunchly Federalist *Centinel* (which had advocated New England's seceding from the Union a few years before) felt compelled to remind his readers that, while the current weather might be unsettling, it should not "excite any distrust of the continued goodness of the God of the Harvest."[4] These words of advice were widely reprinted. (Concern about a

---

[1] Stewart H. Holbrook, *The Yankee Exodus: An Account of Migration from New England* (New York: Macmillan, 1950), 48.

[2] The Puritans' idea of God's omnipotence made it impossible for them to see any natural events as occurring randomly or without some higher purpose. In sermons, ministers admonished their "backsliding" congregations to become more devout as a way to avoid future natural catastrophes. Virginia Anderson, *New England's Generation: The Great Migration and the Formation of Society and Culture in the Seventeenth Century* (Cambridge: Cambridge University Press, 1991), 195.

[3] "Remarkable Weather," *Hampshire Gazette* (Northampton MA), 12 June 1816.

[4] "The Weather," *American Advocate* (Harpswell ME), 15 June 1816. In 1812, Russell had hung a map of Massachusetts over his office, showing the redistricting which Democratic lawmakers in the state had accomplished in order to assure the re-election of the Democratic-Republican governor Elbridge Gerry. When the painter Gilbert Stuart visited Russell's office, he drew a head, wings, and claws around one offending district, commenting, "That will do for a salamander." To this, the quick-witted journalist

crisis of faith was also privately expressed, by diarists such as David Thomas, who accused those who questioned God's wisdom and omnipotence of possessing a "morbid sensibility" and admonished them not to "overlook our blessings."[1]) The *Worcester Gazette* also rebuked this skepticism: "From the North to the South, the same murmurs are heard: the same apparent disposition is manifested to arraign the goodness of Providence."[2] Another editorialist chided "melancholy presages" for predicting doom and gloom. The writer was content to leave the "rule of the seasons" in the hands of God, as recommended in the hymn "Wisdom of Providence":

> Wait, O my soul, thy Maker's will! — Tumultuous passions, all be still!
> Nor let a murmuring thought arise;
> His providence and ways are wise.[3]

This most recent warming trend, with its promise of a second corn crop and bountiful rye, prompted other commentators to wag their fingers at the doubting Thomases, as well as those opportunists who had sought to profit from fears of crop failures: "On the whole, we suspect that the speculator who has been hoarding up grain & crying FAMINE, will gain a LOSS," crowed a correspondent for the *Vermont Gazette*.[4] The clergy were quick to point out how poorly mere mortals understood God's ways. Speculating about causes of the weather only revealed this innate ignorance. For the Harvard-educated Reverend Bentley, a Unitarian presiding over a small Congregational flock in Salem, harebrained theories involving the moon and giant sunspots only demonstrated the "small progress of science and of the advantages nominal science has over superstition & prejudice & ignorance."[5] Bentley blithely predicted the current grousing about cold weather would soon give way to complaints about the heat. The fact that so many of his parishioners in Salem — then New England's pre-eminent seaport — made their living from fishing and overseas trading made it easier for him to downplay the woes of farmers inland. Likewise, Thomas Robbins, searching for an appropriate scriptural passage for his sermon on June 9th — one which might allay the anxiety he had seen on the faces of South Windsor farmers after three "extraordinary" days of frost and frozen cornstalks — chose Jesus' parable of the fig tree, as narrated in Luke:

---

responded, "Better say a Gerrymander." Thus, a new political coinage was born. *Encyclopaedia Britannica*, 11th ed., Vol. 11 (Cambridge: Cambridge University Press, 1910), 904.

[1] Thomas, *Diary*, 56.
[2] "The Season," *Worcester Gazette*, 22 June 1816.
[3] Untitled article, *Ulster Plebian*, 25 June 1816.
[4] "Domestic Summary," *Vermont Gazette*, 2 July 1816.
[5] William Bentley, *The Diary of William Bentley, Pastor of the East Church, Salem, Massachusetts*, vol. 4 (Salem: Essex Institute, 1914), 392.

> A man had planted a fig tree in his vineyard; and he came seeking fruit on it and found none. And he said to the vinedresser, "Lo, these three years I have come seeking fruit on this fig tree, and I find none. Cut it down; why should it use up the ground?" And he answered him, "Let it alone, sir, this year also till I dig about it and put on manure. And if it bears fruit next year, well and good; but if not, you can cut it down."[1]

His message was twofold: first, be patient. The ways of God are not fully known to us, and we must accept whatever bounty He brings, or declines to bring, trying as this may be. Secondly, God's patience with lowly man is not infinite. The chance to repent and embrace eternal life through Jesus will not always be available. We refuse it out our own peril. This remarkable ordeal of snow and ice was a warning. His parishioners must act — or perish. After riding over to East Hartford on Wednesday to attend an ordination, Robbins was pleased by the "concourse of people" which had turned out for this solemn occasion. Their presence in such number suggested that revivalism was, indeed, gaining ground against the forces of wickedness, so recently spreading unbridled, like a plague.[2] His flock was heeding God's message.

But whom was the Almighty testing? It seemed New Englanders were not the only ones enduring unseasonably cold weather. Thanks to the proliferation of newspapers in early nineteenth-century America, people in New England learned that a punishing cold was also afflicting other parts of the country. On June eleventh, the *Connecticut Courant* reported that Virginia had experienced frost the previous Thursday. Much of the Deep South was also suffering from the same drought which was devastating the Northeast. Newspapers ran reports from Quebec, Montreal, and other Canadian cities, describing unprecedented fluctuations between winter-like chills and midsummer heat. In the first week of June, Montreal had sweltered through "hot and sultry" weather one day, and on the next it had snowed.[3] In addition, papers and letters brought by ship across the Atlantic carried news of equally bizarre weather on the Continent. A man in Boston heard from a friend in France that temperatures there had been extremely variable in April, dropping fifty degrees in the course of one day.[4] Reading these reports, some editors in the Northeast realized that sharing weather-related information could be useful in figuring out why these strange and alarming cold snaps were occurring. The *Providence Patriot* took the lead in this early effort at meteorological data-collecting, reprinting accounts from all over

---

[1] Luke 13:6-9, rev. standard version. See Robbins, *Diary*, entry for 9 June 1816, 670.
[2] Robbins, *Diary*, entry for 10 June 1816, 671.
[3] "Extraordinary Season," *Providence Patriot*, 22 June 1816.
[4] Untitled article, *Connecticut Courant*, 11 June 1816.

the Union — from Vermont, Ohio, Pennsylvania and Southern as well as Canadian sources — about exceptionally volatile and crop-destroying weather. [1] These publishers had recognized that heavy rain, drought, high winds, frost, snow, and thunderstorms were not isolated, local events. Larger forces were at play, and they operated across a wide geographical area. This realization had two important consequences. First, it called into question the theological argument that extreme weather phenomena were a sign of God's unhappiness with *particular* persons or groups, in particular places. Secondly, it encouraged close observation and accurate recording of climatic changes. Indeed, this analytical approach to the weather had the effect of transforming drought, earthquakes, and lightning strikes into natural events, governed by forces operating without the controlling hand of God.

Such a radically new understanding of the weather challenged not only orthodox Christian beliefs, but also the premises of 19th-century science. This was the case because religious belief and study of the physical world were so inextricably entwined. Most natural scientists were appalled at the idea of a universe functioning without divine guidance, since their careers were dedicated to revealing this underlying truth. A primary example of this prevailing outlook was Parker Cleaveland, who was teaching at Bowdoin College, in Brunswick, in the District of Maine. The son of a surgeon and grandson of a Puritan minister, he had joined its faculty shortly after graduating from Harvard in 1799. At first Cleaveland had been chiefly interested in reading French and English poetry, but one day stumbled upon his true calling by happenstance. Workmen blasting rock to make way for an aqueduct had come upon some glittering cubes embedded in a granite ledge, which they hoped might be gold. They gave these to Cleaveland to identify. The specimens turned out to be worthless iron pyrite, but he was immediately hooked by this impromptu investigation and began collecting whatever rocks and minerals he could lay his hands on in the surrounding countryside and riverbeds. In the spring of 1808, Cleaveland taught the first courses in natural science ever offered at Bowdoin, in chemistry and mineralogy. Soon he had amassed two of the most complete collections of rocks and minerals then in existence (one his own, the other for the college) and become internationally known in the field of mineralogy, which had scarcely existed in the United States before it became his obsession. In 1816 Cleaveland was about to publish his magnum opus, *An Elementary Treatise on Mineralogy and Geology*, the first book on these subjects in this country. It would earn him words of praise and respect from eminent scientists and thinkers like Goethe and Humphry Davy.[2]

---

[1] "Snow Storm," *Providence Patriot*, 22 June 1816.
[2] Leonard Woods, *Address on the Life and Character of Parker Cleaveland, LL.D.* 2nd ed. (Brunswick ME: Joseph Griffin, 1860), 19, 32, 38, 45.

Over the years Cleaveland had also found time to keep meticulous records of the weather, checking a thermometer outside his house three times a day and measuring the size of unusual phenomena such as hailstones. For over half a century, up until just before his death, in 1858, he filled ledger after ledger with these meteorological data, sharing them with counterparts at other colleges and other scientifically curious parties. One of these was Edward A. Holyoke, the son of a Harvard president, a prominent Massachusetts physician, and the first person in New England to own a thermometer.[1] At his home in Salem, Holyoke jotted down the temperature eight times a day — four times inside, four times outside — for some thirty six years, up until his death at the age of one hundred in 1829.[2] This meticulous recordkeeping was pioneering scientific work. Before Cleaveland began keeping his own ledger, no one in the District of Maine had bothered to record weather information.[3] In fact, in the entire Northeast only a handful of academics were then doing so.[4] In Williamstown, Professor Chester Dewey, a botanist whose thirst for knowledge led him to take up theology, geology, and chemistry as well, was also keeping records.[5] Dewey not only wrote down the temperature every day, but also made observations of the aurora borealis, sunspots, phases of the moon — and the arrival of bluebirds each spring.[6] John Farrar at Harvard, Frederick

---

[1] Holyoke's father, also named Edward, had installed a thermometer in the president's residence by the late 1740s. See Edward A. Holyoke and Enoch Hale, "A Meteorological Journal from the Year 1786 to the Year 1829, Inclusive, with a Prefatory Memoir," *Memoirs of the American Academy of Arts and Sciences* 1(31 January 1833): 107-9. This thermometer was given to him by the Royal Society of London. It used the imprecise Hauksbee scale, which set body temperature at zero degrees. Stommel and Stommel, *Volcano Weather*, 19.

[2] Ibid., 117. In 1789, Holyoke had proposed a new way of calibrating thermometers so that greater extremes of cold and heat could be measured. Under his scale, the temperature at which water froze would be one hundred degrees, and the boiling point 350 degrees. Holyoke, "A Proposal for Adjusting a New Scale to the Mercurial Thermometer," *Memoirs of the American Academy of Arts and Sciences* 3:1 (1809): 54.

[3] Letter of Parker Cleaveland to Levi Hedge, 29 September 1808, reprinted in "Meteorological Observations, Made at Bowdoin College," *Memoirs of the American Academy of Arts and Sciences* 3:1 (1809): 119.

[4] Ezra Stiles began recording temperatures in 1760, at New Haven, using a Fahrenheit thermometer given to him by Benjamin Franklin. Holmes, Abiel, *The Life of Ezra Stiles* (Boston: Thomas and Andrews, 1798), 102. For decades, Yale presidents and professors would rise at 4:30 in the morning to note the temperature. Stommel and Stommel, *Volcano Weather*, 22.

[5] Henry Fowler, *The American Pulpit: Sketches, Biographical and Descriptive, of Living American Preachers, and of the Religious Movements and Distinctive Ideas Which they Represent* (New York: J.M. Fairchild, 1856), 52.

[6] "Weather, Stars, and Living Nature," Chapter Two in "History of Science at Williams." http://www.williams.edu/sciencecenter/center/HistSci00/chapter2.html.

Hall at Middlebury, and Jeremiah Day at Yale made up the rest of this small meteorological fraternity. [1]

Their observations had been made possible by the greater availability of instruments like the barometer, the hygrometer (which measured moisture in the air), and, most significantly, the thermometer. While these devices had been in existence since the time of Galileo, they had not been used to measure atmospheric changes until the second half of the eighteenth century. Several improvements had made the thermometer an accurate and reliable instrument — namely, the adoption of sturdy, weather-proof materials (brass, glass, and steel), better methods of calibration, and the use of mercury.[2] The so-called "Fahrenheit" thermometer enabled academics like Chester Dewey to gauge the rise and fall of the temperature with precision, but this new capability did not embolden them to take the next step of looking for patterns in these changes, or wondering if their observations could help to predict the weather. In this regard they were like other "natural philosophers" of that day, who did not believe it lay within their purview to theorize about the phenomena they were observing. To some extent, this reluctance grew out of a lack of knowledge. Like Isaac Newton before them, they were disinclined to speculate about matters they could not fully measure. In the early nineteenth century, meteorology was still in its infancy. The most these proto-scientists could contribute was the compilation of data, performed with an "almost obsessive urge to pin down every stray morsel of discernible fact and give it permanent recognition in print."[3] More significantly, inquiring into causes meant questioning the ways of God. As devout Christians, professors of natural science were not willing to do this. Instead, collecting facts and specimens was their way of demonstrating that the universe was devised according to an infinitely wise, divine plan. They could not hope to comprehend it. They could only look for God's fingerprints.

June's unsettlingly erratic weather subsided toward the end of the month, and it appeared things were returning to normal. Horatio Seymour, a young man building a house in Middlebury, Vermont, took time out from his labors to write his parents that the temperature there was now pleasant enough that they should consider coming up from Litchfield for a visit. This was a great relief after a remarkably frigid Vermont spring, "perhaps as distressing

---

[1] For Farrar's readings for April and May 1816, see "Meteorological Journals," *North American Review* 3:8 (July 1816): 283-84. Day continued the meteorological observations made by Ezra Stiles during his long tenure at Yale.

[2] Christa Jungnickel and Russell McCormmach, *Cavendish* (Philadelphia: American Philosophical Society, 1996), 161.

[3] Bill Bryson, *At Home: A Short History of Private Life* (New York: Doubleday, 2010), 433.

a time as this part of the country has experienced," Seymour wrote. The last few days of constant sunshine had "strengthened our hopes of the crops that are now growing."[1] Indeed, a thaw was taking place across New England. After having been frozen over the second half of June, the wells in Lyman, in New Hampshire, were providing residents with a regular supply of fresh water.[2] Coastal Connecticut farmers who had recently hoed corn with mittens on now felt their brows glisten with sweat as temperatures reached the low 80s on June nineteenth. In Quebec, the wheat fields and meadows "never looked better than at present," despite the fact that the crops were several weeks behind schedule.[3] On the twenty third and twenty fourth, Albany saw highs in the nineties.[4] In East Windsor, Thomas Robbins took advantage of this warm spell to plant watermelon seeds in his garden.[5] When the Reverend Bentley climbed the stairs to his pulpit in Salem's one-hundred-year-old East Church to deliver a sermon on the twenty third, it was over a hundred degrees, and the oppressive temperature seemed to increase with each upward step he took. He could barely suppress a smile of satisfaction. The weather had improved just as he had predicted it would. Once again, the Lord had provided. He could hear worshippers in the nearby gallery murmuring that it looked like a "fine season for vegetation."[6] A wave of optimism swept across gloomy New England. In Guilford, an otherwise saturnine Amanda Elliott was sufficiently buoyed to enter the succinct phrase "Good weather" in her diary entry on July 3rd.[7]

Then, amazingly, winter rushed back, on — of all days — the Fourth of July. Taking a day off from his clock-making apprenticeship in Plymouth, Chauncey Jerome watched some men playing quoits wearing heavy overcoats, under a midday sun. "A body could not feel very patriotic in such weather," he drily noted.[8] The numbing chill spread across the region and beyond: frost was reported as far away as southern Pennsylvania.[9] At the northern tip of Lake Skaneateles, in the Finger Lakes, people peered out their windows on the morning of the fifth and saw ice glazing the ground.[10] The next day ice

[1] Letter of Horatio Seymour to his parents, Moses and Molly Seymour, 16 June 1816, "Horatio Seymour" folder, Seymour Family Papers, Litchfield Historical Society.
[2] Caleb Emery, "The Cold Summer of 1816," *Collections, Historical and Miscellaneous, and Monthly Literary Journal* (Concord NH: J.B. Moore, 1823), 254.
[3] "The Weather," *Boston Recorder*, 10 July 1816.
[4] "Weather," *Vermont Gazette* (Bennington), 23 July 1816.
[5] Robbins, *Diary*, entry for 24 June 1816, 672.
[6] Quoted in Mussey, "Yankee Chills," 437.
[7] Elliott, "Diary," entry for 3 July 1816.
[8] Jerome, *American Clock Business*, 32.
[9] Untitled article, *Connecticut Mirror* (New Haven), 22 July 1816.
[10] Kihm Winship, "The Cold Summer of 1816." http://home.earthlink.net/~ggsurplus/coldsummer.html. See also Edmund N. Leslie, *Skaneateles: History of its Earliest Settlement and Reminiscences of Later Times* (New York:

formed in ponds across Maine, and on the sixth, seventh, and ninth, it snowed outside Montreal. "Summer snow" fell in Maine and western New York.[1] On the morning of the ninth, Professor Cleaveland's thermometer in Brunswick registered 33.5 degrees.[2] The newly appointed Congregationalist minister in Alstead, New Hampshire, was dismayed to discover ice encrusting the bright green grass outside his house on the same morning.[3] Reports from New Hampshire, Maine, and Vermont all conveyed the same "melancholy" mood, darkened by warnings there might not be enough bread corn to feed families during the coming winter.[4] Many New Hampshire families simply could not afford the high price now being commanded.[5] Anticipating a poor harvest, the administrator of Lower Canada banned the exporting of wheat, beans, peas, barley, and cereal grains, making the situation in neighboring U.S. states even direr.[6] In Vermont, corn had not grown an inch during the last six weeks. Because the grass was frozen, farmers there were forced to feed their cattle vegetables and grain to keep them from starving.[7] A farmer in Kennebec County, in present-day Maine, concluded that, with most of his crops decimated by this latest frost, "a famine for man and beast, seems to stare us in the face."[8] To Vermonters, the recurring snow, the sight of stiff, blackened cornstalks in rock-hard fields, evoked the "most gloomy apprehensions of distressing scarcity" in the months ahead.[9]

In searching for explanations, many instinctively looked to God. An adventuresome Mainer who had ventured west through New Hampshire and Vermont to Utica, New York, finding the corn along his route "miserably pale and puny" and fear of starvation pervasive, reasoned that a Providence which had recently taught Americans a harsh lesson by war was now bent on teaching them one by famine. Mankind's dependence on divine power was brought into sharp relief.[10] The Reverend Robbins discerned God's continuing retribution in this latest destruction of crops and drew upon the words of contrition in Psalms 51:4 ("Against thee, thee only, have I sinned, and done that which is evil in thy sight, so that thou are justified in thy

Andrew H. Kellogg, 1902), 92-3.

[1] "The Season," *Reporter*, 17 July 1816.

[2] Stommel and Stommel, *Volcano Weather*, 37.

[3] Seth Shaler Arnold, "As the Years Pass — The Diaries of Seth Shaler Arnold (1788-1871), A Vermonter," *Vermont History* 8:2 (June 1940): 109.

[4] "The Weather, Etc." *Providence Patriot*, 13 July 1816, quoting from "Frost in July," *Boston Centinel.*

[5] In the spring of 1817, corn sold for as much as $2 a bushel in New Hampshire. Perley, *Historic Storms*, 212.

[6] "The Season," *Boston Recorder*, 24 July 1816.

[7] "The Season," *Newburyport Herald* (Newburyport MA), 19 July 1816.

[8] "Scarcity of Crops," *Boston Gazette*, 15 July 1816.

[9] "The Season," *Reporter*, 17 July 1816.

[10] "By the Mails: The Season," *Bangor Weekly Register*, 3 August 1816.

sentence and blameless in thy judgment") to startle his congregation out of its stubborn spiritual sloth one frosty Sunday.[1] The evil of sin, he reminded them, "consists, principally, in its offense against God — in "rejection of that authority which he justly maintains over all his creatures."[2] As sinners, they must acknowledge their disobedience and submit. Recognizing the limits of their understanding, many New Englanders were inclined to accept what admonishing clergymen like Robbins were telling them, through the parable of the fig tree.[3] Their Maker — the Great First Cause — was causing this weather, as Job's friend Elihu had declared: "By the breath of God frost is given, and the breadth of the waters is restrained."[4] The best thing they could do was to humbly beseech His forgiveness. A workman in South Windsor so earnestly took his minister's words to heart that he refused to chop down a bare fig tree, telling its owner he need only wait until next year — when "the affliction will pass" — for it to bear fruit.[5]

Many others lacked his stubborn confidence. They were plainly frightened. The sense of fear and foreboding then gripping most of New England was evident in a poem about an eerie apparition, published, in mid July, in the New Haven-based *Connecticut Journal*:

> Fair as a snow-ball was its face,
> Like icicles its hair;
> For mantle it appear'd to me
> A sheet of ice to wear.
>
> Though seldom given to alarm,
> I'faith I'll not dissemble,
> My teeth chatter'd in my head,
> And every joint did tremble.

---

[1] Robbins, *Diary*, entries for 6 and 7 July 1816, 673. See also Robbins, "Sermon Preached at East Windsor, July 7th, 1816," in "Robbins, Thomas" folder, Thomas Robbins Papers, Connecticut Historical Society. Robbins also delivered this sermon on 28 July 1816, in Glastonbury, and on 22 December 1816, in Hartford.

[2] A poem published in 1816 well captured the fear instilled in many by what they regarded as God's anger at their sins: If God withholds those milder rays / And sends us frosts & chilling days / E'en snow as late as eight of June / That nips the fruit in early bloom / Shall we be frightened by such things? / No, rather frightened at our sins. T.D. Seymour Bassett, "Fitch against Emigration: Hyde Park, Vermont, 1816," *New York Folklore Quarterly* 28 (March 1972), 39. Quoted in Michael Sherman, Gene Sessions, and P. Jeffrey Potash, *Freedom and Unity: A History of Vermont* (Barre VT: Vermont Historical Society, 2004), 166.

[3] Robbins used this text again for a sermon delivered on 4 August.

[4] Job 37:10. Quoted in "The Season," *Reporter*, 17 July 1816.

[5] Brian Fagan, *The Little Ice Age: How Climate Made History, 1300–1850* (New York: Basic Books, 2000), 175.

> At last I cryed [sic], "Pray who are you,
> And whither do you go?"
> Me thought the phantom thus replied —
> "My name is Sally Snow."[1]

At that point in the month, ice still covered many small lakes in Quebec, and it remained bitterly cold along the New York shore of Lake Erie. But, further south, the air was warming again.[2] Thomas Robbins plucked enough pea pods for a couple of evening meals on the eleventh. This small gift from God came with other rewards: several members of his community had "got religion" thus far this year, he was pleased to see.[3] Their conversion was part of a major revivalist movement then taking place in the United States — given additional impetus, no doubt, by the adverse weather.[4] As Benjamin M. Palmer, a Congregational preacher in Charleston, South Carolina, related in one of his sermons, conversions were increasing, and the growing interest in religion was "unquestionably . . . one of the most prominent signs of the times."[5] The rest of July and the first weeks of August were reassuringly pleasant. The ice melted away, crops were reseeded, and new shoots soon broke through the soil promisingly, while, in some areas, the drought which had been afflicting the entire Atlantic seaboard for months was broken. Abundant showers in mid-August quieted talk of crop failures and famine and revived hopes for the rye and barley harvest. A bumper wheat crop was now forecast for the Mid-Atlantic and Southern states, and fruits appeared "highly promising," since they had been spared the ravages of pests this year. Young corn in New York was also said to be "wonderfully improved" over this five-week period.[6] An editorial in the *New Hampshire Patriot* proclaimed a complete reversal of fortune: "the prospect of a general famine has succeeded one of uncommon plenty."[7]

---

[1] "The Apparition," *Connecticut Journal*, 16 July 1816.

[2] Sutherland, "June 1816 Snowfalls," 13.

[3] Robbins, *Diary*, entry for 12 July 1816, 673.

[4] At its annual meeting, on June eighteenth, the Connecticut association of Congregational ministers had hailed this trend: a year in which so many had entered the church "has probably never been witnessed in this country," proclaimed one report on this Hartford conference. "Very Interesting Religious Intelligence," *Farmer's Cabinet* (Amherst NH), 10 August 1816. In northern Litchfield County alone, congregations had added some twelve hundred new members in the previous year. "State of Religion in Connecticut," *Boston Recorder*, 10 July 1816. In Salisbury, 172 persons joined the local church despite the absence of a minister. Charles Roy Keller, *The Second Great Awakening in Connecticut* (New York: Archon, 1968), 46.

[5] Benjamin M. Palmer, *Signs of the Times Discerned and Improved: in Two Sermons Delivered in the Independent or Congregationalist Church in Charleston, S.C.* (Charleston: J. Hoff, 1816), 7.

[6] "The Season," *Yankee* (Boston), 19 July 1816.

[7] Quoted in Heidorn, "Eighteen Hundred and Froze to Death."

Around Boston, these same showers came rise to "joyful thanksgiving and praise," with fruits, vegetables, grasses, and grains now in bountiful supply.[1] Further up the Massachusetts coast, the scare caused by corn shortages abated, after prices had dropped back a quarter to $1.25 a bushel on the Boston market. Assessing Salem's still-robust economy, the Reverend Bentley chided alarmists for having overreacted to the spell of bad weather. "The agricultural prospects are as good as ever, after all our complaints," he asserted. Moreover, sixteen fishing vessels returning from Newfoundland's Grand Banks had just brought in a catch of half a million cod.[2] For Bentley, this boded a Panglossian "best of all possible worlds." On the last day of July, his fellow cleric Thomas Robbins was likewise delighted to record that the "harvest comes in very well," after admitting just a few days before that the grain was hardly fit to be cut.[3] As a precaution, however, some editorialists were now urging farmers to plant potatoes, in case cold weather returned. (Potatoes, which had first been introduced in the Northeast by Irish immigrants about a hundred years earlier, could survive light frosts.) In the Mid-Atlantic states a bigger than usual crop was helping to offset the damage done earlier by frost to "Indian corn."[4]

But the rains did not last. Instead, the dry spell which had started back in late April, affecting states from Maine to Ohio and Virginia, worsened. There were no more downpours. Some days it sprinkled for a few minutes, but despite this the ground cracked and turned to lifeless dust. Some places had no rain at all: Vermont did not see a drop for a stretch of 120 days — until September, when it was too late to do any good. Lightning sparked forest fires across that state, parts of western New York, and in Maine and New Hampshire. Flames burned uncontrollably for weeks, destroying homes and barns, filling the sky with acrid, choking smoke, and making an already dismal season even more lugubrious.[5] Over one thousand acres were incinerated near Providence.[6] The Connecticut River sunk to its lowest level in forty years.[7] Whatever little corn had managed to survive the frost

---

[1] "The Season," *Providence Patriot*, 10 August 1816.
[2] Bentley, *Diary*, 403, 407.
[3] Robbins, *Diary*, entries for 26 July and 31 July 1816, 675.
[4] Untitled article, reprinted from the *Gazette of the United States, Connecticut Journal*, 16 August 1816.
[5] "Boston, October 15," *Providence Patriot*, 19 October 1816. Smoke on the Kennebec River was so thick that ferry pilots had to rely on compasses to navigate across. See "The Season," *Providence Gazette*, 19 October 1816. In western Vermont, visibility on roads was reported to be only twelve feet. Michael Sherman, Gene Sessions, and P. Jeffrey Potash, *Freedom and Unity: A History of Vermont* (Barre VT: Vermont Historical Society, 2004), 166.
[6] Perley, *Historic Storms*, 206.
[7] So reported several New England newspapers. See, for example, "The Season," *Dedham Gazette* (Dedham NH), 6 September 1816.

just a few weeks before was now endangered. Hay shortages in northern New England once again fueled fears of a severe shortfall of livestock feed. Surveying the situation from his home in Monticello, 73-year-old Thomas Jefferson penned a letter to his friend and treasury secretary Albert Gallatin, then serving as U.S. Minister to France, informing him: "We have had the most extraordinary year of drought and cold ever known in the history of America..."[1] The James River was shallower than at any point since the first English settlers had arrived there, in 1607 — and still falling. Virginia had been spared damage done by the extraordinary frosts of June and July, but the ones in August, combined with the continuing drought, meant that the corn crop would be only a third its usual size, and tobacco, backbone of the state's economy, would be even less, and "of mean quality."[2]

In northernmost New England, people were beginning to despair for their future. The growing season was turning out to be extremely short, and they would have little to show for it. Their fears were well grounded. Along the coast of Maine, instead of the usual 120 days, there would be only sixty warm enough for crops to grow.[3] In Connecticut, there would be fewer than seventy.[4] For the region as a whole, 1816 would end up having the most abbreviated growing season on record. Panic over anticipated winter shortages drove up prices: agents combed markets from New York to Canada looking for flour — and paying more than three times the normal price for a barrel. Speculators bought up all they could and hoarded it, hoping to turn a large profit at the end of the season.[5] Fearing a massive crop failure and a resulting famine, many congregations came together to pray and fast. Fasting was an age-old method of persuading God to let it rain. In the New World, the Pilgrims had first resorted to fasting in 1623, to end a long drought which threatened starvation. After singing hymns, praying for God's mercy, and going without any food for over eight hours in their meeting house, these

---

[1] Letter of Jefferson to Gallatin, 8 September 1816. Jefferson, *The Writings of Thomas Jefferson*, vol. 10, *1816-1826*, ed. Paul Leicester Ford (New York: G.P. Putnam's Sons, 1896), 64-5.
[2] Thomas Jefferson, *The Works of Thomas Jefferson*, vol. 12, ed. Paul Leicester Ford (New York: G.P. Putnam's Sons, 1905), 37. See also "On the Climate," *National Daily Intelligencer*, 27 August 1816.
[3] William R. Baron, "1816 in Perspective: The View from the Northeastern United States," in *The Year Without a Summer? World Climate in 1816*, ed. C. R. Harington (Ottawa: Canadian Museum of Science, 1992), 133. Cf. Clive Oppenheimer, "Climatic, Environmental and Human Consequences of the Largest Known Historic Eruption: Tambora Volcano (Indonesia) 1815," *Progress in Physical Geography* 27:2 (2003): 245: Fig. 8: "Length of Growing Season in (a) Southern Maine, (b) southern New Hampshire, and (c) eastern Massachusetts, between 1790 and 1840." Oppenheimer puts the Maine growing season at seventy days for 1816.
[4] Stommel and Stommel, *Volcano Weather*, 42.
[5] Skeen, *America Rising*, 6.

worshippers were rewarded by abundant showers, which revived their crops and saved the colony.[1] So it is not surprising that ministers all over New England now turned to this remedy. In the crossroads village of Benson, in western Vermont, a large portion of the fifteen-hundred-odd population showed up at the Church of Christ on Wednesday, July seventeenth, to beg for forgiveness and an end to God's anger. Pastor Dan Kent, a charismatic, seasoned speaker, began this day of fasting and worship with a "fervent and animated" prayer: "Ask and ye shall receive." As the service progressed, black clouds welled up like bruises on the horizon, and when the minister rose again, to read a passage from the Bible ("And it shall come to pass, that before they call, I will answer; and while they are yet speaking, I will hear"), "the torrents of rain, which poured down from the clouds, rendered his agitated voice almost inaudible." The downpour was so intense that it kept the congregation from leaving the church. Finally, the drops slowed to a trickle, and intense rays of light beamed from the west just as the day was drawing to an end. God's grace was manifest.[2]

Twice during August the Reverend Robbins implored God to end the drought which was destroying the corn and turning central Connecticut into a dustbowl.[3] On August eighteenth, a particularly stifling day, he summoned his parishioners to a prayer service. During this gathering, a few drops fell, but they did little to save the withering crops. Isolated showers brought some relief to neighboring towns on the twentieth, but skirted East Windsor. On the evening of the twenty fifth, Robbins tried again, convening another prayer meeting. This time, the imprecations appeared to succeed, as a "moderate and very refreshing" rain fell the next day, saturating the dust and giving him reason to offer thanks. It even rained a bit the following day.[4] Other parts of New England weren't so lucky. Chester Dewey's gauge at Williamstown would accumulate only an inch-and-a-third of rainwater from the last week of August to mid-October.[5]

Of course, not all prayers evoked a response from the heavens. It is not known how many ministers and public officials held special services to propitiate a displeased God during the summer of 1816, but, given the fact that it rained so little and so infrequently, one can surmise that many churchgoers would have felt their prayers had been in vain. For some unfathomable reason, the Almighty still remained unhappy with them. Why was this so? Why was God withholding the "dews of divine grace" from their fields, orchards, and vineyards?[6]

---

[1] James W. Baker, *Thanksgiving: The Biography of an American Holiday* (Lebanon NH: University of New Hampshire Press, 2009), 24.
[2] Untitled article, datelined Benson, 26 July 1816, *Vermont Mirror* (Bennington), 7 August 1816.
[3] Robbins, *Diary*, entry for 20 July 1816, 674.
[4] Ibid., entries for 18, 25, and 26 August 1816, 676-7.
[5] Dewey, "Meteorological Observations," 156.
[6] "Domestic Intelligence," *Connecticut* Journal, 9 July 1816.

As if summer temperatures in the thirties, snow, and drought weren't sufficient punishment, the coup de grace was delivered in the second half of August. On the thirteenth, temperatures in Rhode Island plummeted from the mid-eighties to the thirties within a few hours as a "boisterous and cold" wind swept down from the north. Frost — frost in August! — draped New England fields like a gauze curtain.[1] From Boston to Stockbridge in the Berkshires, Massachusetts fields sparkled with an unseasonal, hoary white.[2] So far, frost had had been spotted in each month somewhere in New England, and it seemed likely that it would reappear as regularly during the rest of the year. It was this sobering realization which would give rise to the sobriquet "year without a summer" for 1816, even though this was a misnomer. The region had had brief spells of hot weather; it was the dramatic oscillations between extreme heat and extreme cold which made the year so singular. These premature frosts badly damaged most of the corn which farmers had optimistically planted during the short hot period a few weeks before — their third attempt to sow this crop. The wintry chill spread to Maine, the counties around Boston, New Hampshire, and Vermont, where it killed off potatoes, beans, and corn on the twenty first. [3] Only a blanket of night fog spared corn growing along the Connecticut River.[4] Snow dusted the White Mountains. Down south, in the Carolinas, the cold snap injured tobacco, corn, and cotton plants. Toward the end of August, ice was observed floating on the Potomac River a dozen miles from the White House.[5] To the west, in Kentucky, grapevines wilted. [6] Although it rained in the Ohio Valley, the summer-long drought continued in the East. Wells in Maine were as low as they had ever been.[7] After sampling his first ears of green corn and complaining about the "severe" heat on the nineteenth of August, Thomas Robbins was distressed to see the temperature fall sharply a few days later and then, on the twenty ninth, to find a new sprinkling of frost on the ground. Overnight, the look of the landscape had turned "melancholy," as it looked in western Massachusetts when he visited Lenox and Pittsfield the following day. Although the wheat, oats, and rye there appeared unaffected, much corn was lost. A week later, after returning the sixty miles from Lenox on horseback, Robbins was unusually pessimistic when he sat down at his writing table and took up his pen: "I presume no person living has known so

---

[1] "The Season," *Providence Patriot*, 24 August 1816.
[2] Perley, *Historic Storms*, 210.
[3] Lesley-Ann Dupigny-Giroux, "Climate Variability and Socioeconomic Consequences of Vermont's Natural Hazards: A Historical Perspective," *Vermont History* 70 (Winter/Spring 2002): 37.
[4] Perley, *Historic Storms*, quoted in Stommel and Stommel, *Volcano Weather*, 40.
[5] "A.B.," "The Season," *American* (Hanover NH), 28 August 1816.
[6] "Washington (Ken.), Aug. 30," *National Daily Intelligencer*, 9 September 1816.
[7] "The Season," *Providence Gazette*, 19 October 1816.

poor a crop of corn in New England, at this season, as now."[1]

These ruined harvests became a topic of national concern. It would be the first matter taken up by James Madison in his eighth — and final — address to both houses of Congress, on December 3rd. "In reviewing the present state of our country," the President would remark, "our attention can not [sic] be withheld from the effect produced by peculiar seasons which have very generally impaired the annual gifts of the earth and threatened scarcity in particular districts." Madison would go on to reassure lawmakers that the country as a whole did not face grave food shortages.[2]

With this disastrous growing season now coming to an end, Americans began to try to make sense of it. Most of the United States had just endured the coldest summer anyone could remember. Because records had only been kept for a few decades, it was impossible to know if this was a once-in-a-hundred-year event or a periodically recurring aberration. In the early nineteenth century, people did not imagine long-term climatic cycles. The Christian world view held that the earth was only six thousand years old and had not evolved since its creation. It was kept that way by immutable laws devised by the Creator. But perceptions of the relationship between the natural and supernatural worlds were beginning to change. Close examination of terrestrial objects tended to undermine belief in the story told in Genesis. For example, two decades hence, the Swiss-born paleontologist Louis Agassiz would propose that the planet had gone through ice ages in a distant past, based, in part on his studying alpine rocks that had been moved by glaciers. Such physical evidence acquired an irrefutable reality, separate from religious teachings, if not in outright contradiction to them. This decoupling of material fact from "theological speculations" had, in fact, started in the seventeenth century, with Francis Bacon's development of the scientific method and the appearance of philosophical works like Descartes' *Discourse on the Method* (1637) and Thomas Hobbes's *Leviathan* (1651).[3] These two approaches to terrestrial events like earthquakes, volcanic eruptions, storms, lightning, and drought were not mutually exclusive, but complementary. One could be a devout Congregationalist and still wonder why the barometer fell before it rained. A frontal clash between science and Scripture would not take place until the publication of Charles Darwin's *On the Origin of Species*. What *had* developed by 1816 was another way of seeking

---

[1] Robbins, *Diary*, entries for 29 and 30 August and 5 September, 1816, 677-78.

[2] James Madison, remarks before the House and Senate, 3 December 1816, *The Writings of James Madison*, vol. 8 (New York, London: Gaillard Hunt, 1908), 375-6.

[3] On the subject of the "Great Separation" of religious thought from political philosophy, see Mark Lilla, *The Stillborn God: Religion, God, and the Modern World* (New York: Knopf, 2007).

truth. The churches no longer had a monopoly on this undertaking. This fact alone was unsettling to the devout.

They were perplexed, but clung to their belief in God's benevolent control over their lives. Beneath the suffering and panic caused by failed crops, they searched for a silver lining. They hailed the few good yields that did materialize, the munificence which, in the end, saved the day. Typical of this point of view was an article in the Hanover, New Hampshire, *American*, at the end of August. It conceded this growing season had been trying: "Drought and cold seem to have conspired to blast, if it could be blasted, our confidence in the promise of God, that seed-time and harvest, while the world stands, shall never cease." But, now, as the grain tonnage was turning out to be larger than expected, and a seasonably warm fall augured a return to normalcy, this editor sensed that fears were ebbing: the worst had not come to pass. In the end, trust in God had been justified. Now believers were "led to acknowledge and to bless Him, whose 'tender mercies' are over all his works; Him, who will yet 'open his right hand, and satisfy the desires of every living thing.'"[1] Like ministers offering consoling words from their pulpits, these writers sought to reassure their readers that God was, indeed, in His kingdom, and all was right in the world. One prominent figure making this case was Charles Hosmer, publisher of the *Connecticut Mirror*, a paper he had founded in Hartford a few years previously.

For Hosmer, publishing was a way of proselytizing. Before going into newspapers, he had brought out a version of the Bible containing evangelical commentaries. In 1808, he had started the *Litchfield Gazette* with the primary objective of winning over the people of northwest Connecticut to the conservative, Federalist camp.[2] Once that goal had been reached, he shut the paper down. When his *Connecticut Mirror* took up the subject of the recent calamitous weather, it similarly did so with an ideological intent. In this regard Hosmer was hardly alone. Most newspapers of that day had a clear agenda and made no attempt to hide it. Often their perspective wove together religious and political perspectives. On the fifth of August, the *Mirror* ran an article on "The Season." It began with the quotation from Genesis on God's promise to always provide "seed time and harvest." It acknowledged that many now doubted this promise. These pessimists were predicting seven years of starvation, as had occurred in ancient Egypt. Other papers were printing alarmist pieces about grain shortages, spreading "famine fever" in Connecticut. The *Mirror* article ridiculed these reports and argued they

---

[1] "The Season," *American*, 28 August 1816.

[2] Hosmer had declared that his new paper would be "zealously devoted to the Federalist politics of the nation and to the great essential interests of the Commercial States." Quoted in "Early American Newspapers: Selected Newspaper Descriptions by State," 2. http://www.newsbank.com/readex/PDF/EANMicro%20Selected%20Descriptions.pdf.

should not be printed. It confidently predicted that the "present warm weather" would save the fall harvest. Instead of whining about not having enough food, people should be content with what they had. Complaints and worries about a poor harvest were simply playing into the hands of speculators, who were buying up all available flour and grain and driving up prices. What was needed during this unsettling time was not rash action, but renewed faith in the Almighty.

The people of Connecticut, the "land of steady habits," ought to honor the long-established order of their forefathers. True security, happiness, and prosperity lay in hewing to the tried-and-true ways of the past. Temporary crises and distress did not justify calling these precepts into question. Discontent and dissension were both pointless and harmful.[1] To convince readers the crisis was passing, the same issue of the *Mirror* reprinted articles from papers in New Hampshire and New York, describing recent rainfall there and the promise of abundant food and renewed trust in Providence it had brought with it.[2]

Submission to the Almighty's will was necessary, too. This was a recurrent theme in sermons all over New England that summer. For ministers believed that their congregations' turning away from God had provoked His display of anger in the skies. This fear-laden message could persuade non-believers to return to the fold. If they renounced their "wicked" ways, God would be kindly disposed toward them, as the Reverend Noah Porter reminded his Farmington, Connecticut, parishioners, on August twenty fifth. The Day of Judgment was fast approaching: some would go to Heaven, others to Hell. Their destiny lay in their own hands: those who "cast out demons" and performed good works would be saved.[3] After the return of rain on the twenty sixth, Thomas Robbins preached in nearby Enfield, drawing upon a passage from Revelations to make this point: "Awake, and strengthen what remains and is on the point of death, for I have not found your works perfect in the sight of my God."[4] Speaking the next month he portrayed the continuing cold weather and its destruction of crops as a spiritual test. Frost killed some corn, while sparing the rest. So would it be with the souls of men. The Almighty was unforgiving of sinners who turned away from Him and "cut" them off, but those who remained in His good graces would live

---

[1] "The Season," *Connecticut Mirror*, 5 August 1816.

[2] Untitled article, reprinted from *Utica* (NY) *Patriot, Connecticut Mirror*, 5 August 1816; untitled article from *Rhode-Island American, Connecticut Mirror*, 5 August 1816.

[3] Sermon of Noah Porter, 25 August 1816, "Connecticut Sermons: Noah Porter, Manuscripts," vol. 1, "Farmington, 1802-1821," History and Genealogy Unit, Connecticut State Library.

[4] Revelations 3:2. See Robbins, *Diary*, entry for 27 August 1816, 677.

forever.[1] There were many such "wicked" persons in Robbins' neighborhood: stubbornly, they maintained a "deep enmity against the church."[2] Perhaps now they would change their minds.

But these appeals to patience and faith did not dissuade those keenly curious about the summer's abrupt rises and falls in temperature — sometimes called "Weatherwisers" — from collecting weather-related data and sharing these with their friends and the public.[3] On August twenty first, an Albany man wrote to the editor of the *Columbian*, reporting an overnight plunge in temperature to forty five degrees. This had come after a violent "thunder gust" blowing in from the west, accompanied by hail and a great deal of rain. Frost the following morning had vanished under a light breeze and a seasonal seventy two degrees. Then the temperature had nosedived again, down to forty nine degrees, giving plants a "severe shock." The only (whimsical) explanation the writer could offer for these dramatic fluctuations was that, since the cold winds had originated in Canada, they must be somehow connected with America's "just and necessary" conflict there — that is, the War of 1812. It was clear, the anonymous author mused, that the United States had to annex Canada or else New Englanders would be forced to migrate south to escape this unwelcome cold.[4]

Serious scientific theories were advanced, but they were not grounded in a deep understanding of what causes the weather. In some cases, these notions did not jibe with what had actually happened during 1816. For instance, some educated persons accepted Benjamin Franklin's hypothesis that clearing the forests in North America had exposed them to more sunlight.[5] But this process should have resulted in warmer temperatures, not colder ones. (Herschel's theory that the presence of numerous sunspots allowed more solar heat to reach the earth was likewise contradicted by the prevalence of these black dots during this year of extraordinarily cold temperatures.) The best argument that defenders of Franklin could muster was that the past year was an exception in a general warming trend, already evident along the same latitude in Europe.[6] Weather watchers had to take measurements over several years to discern what was actually occurring.[7]

---

[1] Ibid., entry for 28 September 1816, 681.
[2] Ibid., entry for 19 September 1816, 680.
[3] "The Season," *Connecticut Courant*, 3 September 1816.
[4] "The Season and the Climate," *National Daily Intelligencer*, 3 September 1816.
[5] This theory would be later championed, with some alterations, by Noah Webster. He believed that clearing forests made the country more prone to shorter winters and more variable temperatures. See Webster, *Notes on the Life of Daniel Webster*, vol. 1, ed. Emily E.F. Skeel (New York: privately printed, 1912), 510.
[6] See, for example, "Kennebeck River," *North American Review* 9:3 (September 1816), 324.
[7] "On the Climate," reprinted from the *Richmond Compiler* in the *National Daily*

But this long-range approach did not give New Englanders much peace of mind. They urgently needed to know why their climate had suddenly become so harmful to their crops. More than the freezing temperatures, the blighted fruits and vegetables, and the uncertainty of survival through the upcoming winter, it was the erratic swings in temperature which had unnerved them. As a writer for Boston's *Centinel* put it, "It is certain, more sudden and extensive changes of the weather from hot to cold is [*sic*] not recollected by our oldest citizens than have occurred in the passing summer."[1] Or, in the words of a fellow Massachusetts correspondent, the seasons seemed to have "blended together."[2] Some tried to make light of this, writing poems about the peculiar weather:

> The season, 'tis granted, is very gay,
> But we cannot in justice complain of the weather;
> For if changes delight, we have in one day,
> Spring, Summer and Autumn, and Winter together.[3]

Laugh as one might, this state was psychologically unsettling. Like all living creatures, human beings depend upon regularity in the climate for their physical survival — warmth and a food supply. But predictable seasonal change also gives humans a sense of inner security: things are unfolding as they should. Unusual volatility of the weather can upset this feeling of well-being and make it seem that the world is out of control. This perception, in turn, can create a sense of helplessness, even depression. As newspaper accounts from 1816 indicate, this was clearly the case for many New Englanders. Even though they lived in a region known for its bracing winters, they had always counted on this season giving way to a rejuvenating spring and fruitful summer. But not this year. Their anxiety was amplified because this assumption about seasonal progression had been grounded in Christian belief. Psychological insecurity thus engendered a crisis of faith. Some New Englanders did what their ministers told them to do — to accept the mysterious ways of the Almighty. "What the issue may be, we dare not predict," one Waltham man wrote in September, "but at all events, it becomes us not to complain, as we are assured by unerring Wisdom, that Summer and Winter, seed-time and harvest, shall not fail."[4] For other New Englanders, however, it was not so easy to discard their misgivings. Feeling "abandoned" by God, they could not help examining their relationship with

Intelligencer, 27 August 1816.
[1] "The Season," reprinted from the *Columbian Centinel* in the *Connecticut Courant*, 3 September 1816.
[2] "Boston, September 4," *Connecticut Herald* (New London), 10 September 1816.
[3] "The Weather," *Providence Patriot*, 26 October 1816.
[4] "Boston, September 4," *Connecticut Herald*, 10 September 1816.

Him and worrying about their future.

Farmers had one last chance to salvage a harvest before the winter set in. But the elements continued to conspire against them. The early frosts in the second half of August proved to be harbingers of even colder weather. Snow softened the peaks of Vermont's Green Mountains the first week of September. Two inches fell on Springfield, Massachusetts, on the 10th — a time when ears of Indian corn, McIntosh and Cortland apples, peaches, peas, and cucumbers were usually picked.[1] It seemed "that the polar circle had slipped to the tropic, making of the temperate zone the frigid."[2] Communities across the state held more prayer meetings, but the only concrete result was the flaring up of fires in the still dry forests once the snow had melted.[3] Hard frosts robbed northern New England of most surviving crops. At Dartmouth College, sunrise temperatures in late September reached only the low twenties. In nearby Salem, New Hampshire, corncobs turned into icy lumps and red apples into snowballs.[4] After Lyman Beecher's 41-year-old wife Roxana, mother of Harriet Beecher Stowe, succumbed to tuberculosis in Litchfield on September thirteenth, the ground in the cemetery was frozen so deep that it was difficult to dig her grave. Frost "singed" pumpkin and potato leaves as far south as Asheville, North Carolina.[5] In upstate New York, several nights of hard frost killed off most remaining corn.[6] New England lost virtually its entire crop after another severe frost on the 28th.[7] In coastal Brunswick, Parker Cleaveland was taken aback to find two inches of fresh snow covering his doorstep on October sixth — scarcely ninety days since the spring's last flakes had fallen on this spot.[8] With the fall weather already so intemperate, many envisioned an exceptionally long, hard winter. Ominously, a major snowstorm dumped as much as ten inches on northern New York counties on the seventeenth and eighteenth of October.[9] One elderly farmer was so despondent about not having enough food that he

---

[1] Perley, *Historic Storms*, 211. Farmers in Vermont were similarly prevented from harvesting ears of corn by an unexpected frost on 10 September. Walter Hill Crockett, *Vermont: Green Mountain State.* vol. 3. (New York: Century History, 1923), 134.

[2] *History of Concord, New Hampshire, from the Original Grant in Seventeen Hundred and Twenty-Five to the Opening of the Twentieth Century*, ed. James O. Lyford (Concord NH: Rumford Press, 1903), 351.

[3] Mussey, "Yankee Chills," 443.

[4] Ibid., 444.

[5] "Summary," *Connecticut Mirror*, 14 October 1816.

[6] Untitled article from *Buffalo Gazette*, 1 October 1816, reprinted in *Centinel of Freedom* (Newark NJ), 16 October 1816.

[7] So noted Leonard Hill in his journal. Hill, *Meteorological and Chronological Register*, 65.

[8] "The Season and the Climate," *Providence Gazette*, 19 October 1816.

[9] Sutherland, "June 1816 Snowfalls," 15.

slaughtered all of his cattle and then hanged himself.[1] Adino Brackett, a farmer in Lancaster, New Hampshire, scribbled in his diary: "This past summer and fall have been so cold and miserable that I have from despair kept no account of the weather. It could have been nothing but a repeatation [*sic*] of frost and drought."[2]

Others looked on the bright side: thanks to the lack of rain, wheat and other grains were doing well.[3] These crops might make up for the lost corn, in bread and porridge.[4] Displaying their famed Yankee resourcefulness, farmers also turned to the lowly potato — a relatively new addition to the region's diet and generally denigrated as a poor man's food.[5] Because it grew underground, it was safe from frost as well as foraging pigs. Early in the summer, many farmers had elected to plant potatoes in case their other crops failed, and this decision had turned out to be prescient.[6] At the start of 1817 the Reverend Bentley noted that the potatoes in his Salem garden had experienced "super-abundant growth."[7] Mainers filled their plates with boiled ones and declared them eminently edible. Vermonters figured out more palatable ways to cook them, such as baking in Dutch ovens.[8] Other New Englanders experimented with foods they would previously not have deigned to try. They boiled nettles, leeks, clover heads, and turnips and made stew from porcupines and wild pigeons.[9] In the Green Mountain State, men

---

[1] Maurice Morley, "History Lesson: The Year There was No Summer in Ballston Spa, Milton, Malta and Ballston," 15 August 2010. http://www.saratogian.com/articles/2010/08/15/bspalife/doc4c654a1e7718a615390096.txt?viewmode=fullstory.

[2] Quoted in Deke Rivers, "1816 Volcanoes Made Summer Feel like Winter in Eastern United States," 20 April 2010. http://dekerivers.wordpress.com/2010/04/20/1816-volcanoes-made-summer-feel-like-winter-in-eastern-united-states/.

[3] Untitled article, *Connecticut Courant*, 1 October 1816.

[4] Some corn was being imported from the Ohio Valley, but at the exorbitant price of three dollars a bushel — three times what it customarily cost — this was out of reach for all but wealthy families. As one wag summed up, "We cannot get it, if we could pay for it — and we cannot pay for it, if we could get it." Quoted in "Franklin," "Security of Corn," *Hallowell Gazette*, 23 October 1816. A few desperate men reportedly walked forty miles to buy half a bushel for their emaciated families, paying two dollars for this small portion. Harlan Hatcher, *The Western Reserve: The Story of New Connecticut in Ohio* (New York: Bobbs-Merrill, 1949), 70.

[5] By August, a potato crop "several fold larger" than any previous ones was being harvested in the Mid-Atlantic states. See untitled article from the *Gazette of the United States*, reprinted in *Connecticut Journal*, 6 August 1816.

[6] "The Season," *The Yankee* (Boston), 16 July 1816. See also untitled article, *Connecticut Herald*, 10 September 1816.

[7] Bentley, *Diary*, entry for 24 January 1817, 430. Thomas Robbins considered his potato harvest in September 1816 a "special blessing." Robbins, *Diary*, entry for 5 September 1816, 678.

[8] "Franklin," "Security of Corn," *Hallowell Gazette*, 23 October 1816.

[9] Holbrook, *Yankee* Exodus, 77. See also Crockett, *Green Mountain State*, 135.

---

hurled nets on rivers to catch fish; some fishermen traded this lucrative catch for maple syrup.[1] Upstate New Yorkers foraged for berries and roots in the surrounding forest.[2] In the District of Maine, many families survived by shooting deer and other game.[3] To keep their cows from starving without winter hay, some farmers cured cornstalks as feed.[4] Families shared loaves of bread and other provisions, meager as they were, with hungry neighbors.[5] In northern New England that fall, it was all for one, and one for all.

As the strange year of 1816 came to a close, God-fearing Yankees had good reason to feel they had endured one of the worst growing seasons ever. Indeed, historians would characterize its temperatures as "among the lowest in the recorded meteorological history of the Western world."[6] New England had borne the brunt of this frigid spring and summer — but it had affected much of the country as well as northern Europe.[7] Nationally, temperatures in May had dipped four-and-a-half degrees below the historical mean, in June, six degrees, and in July, ten — the lowest deviation ever recorded for that month.[8] Many in the region tried to shrug off this "year without a summer" as a fluke. It portended no long-term climate change. Others were moved to become more religious, to acknowledge that the harsh weather of 1816 had been a deserved punishment for not upholding God's Commandments. They eagerly joined the revivalist movement then gaining strength across the United States.[9] Orthodox men of faith like Thomas Robbins and Lyman Beecher had cause to rejoice.[10] At the same time, instruments for observing the heavens and gauging terrestrial weather were generating more accurate data. Natural events, like the appearance of sunspots, could now be analyzed in terms of physical forces. More intense investigation raised a profound question: Were these phenomena willed by God or determined by fixed natural laws? Some observers conceded the nation's weather might be undergoing a dramatic, lasting cooling, and that Americans would somehow

---

[1] Mussey, "Yankee Chills," 442.

[2] Winship, "Cold Summer of 1816."

[3] Quoted in Schlegel, "The Year Without a Summer: 1816, in Maine."

[4] "Buffalo, October 1," *Centinel of Freedom*, 15 October 1816.

[5] Crockett, *Green Mountain State*, 135.

[6] John D. Post, *The Last Great Subsistence Crisis in the Western World* (Baltimore: Johns Hopkins University Press, 1977), 1.

[7] It was also the second coldest summer on record for Philadelphia. H.E. Landsberg and J.M. Albert, "The Summer of 1816 and Volcanism," *Weatherwise* 27:2 (April 1974): 66.

[8] "Spots on the Sun," 286. See also Post, *Last Great Subsistence Crisis*, 12.

[9] Winship, "Cold Summer of 1816." This new interest in religion had been evident before the temperatures fell below normal that spring.

[10] Keller, *Second Great Awakening*, 55. For evidence of Robbins' reaction to this religious revivalism, see "Robbins, Thomas Papers, 1816-1822," Folder 1, "Jan 15–Jan 27 1816," Ms. Stack, Connecticut Historical Society, Hartford.

have to adjust to a more taxing environment.[1] Adumbrating Darwin's theory of adaptation, a writer for *Niles' Weekly Register* ruminated: "Vegetables receive new constitutions when transplanted to an uncongenial soil or climate, so will the habitude of our bodies be doubtlessly changed to suit the changes of the seasons."[2] But residents of New England could not easily countenance such a transition. Their immediate concern was getting through the coming winter and then figuring out if they could endure another wretched growing season like this one.

---

[1] Stephen L. Vigilante, "Eighteen-Hundred-and-Froze-to-Death," in *Mischief in the Mountains*, ed. Walter R. Hard, Jr. and Janet C. Greene (Montpelier: Vermont Life Magazine, 1970), 101.
[2] "Climate of the United States," *New-Jersey Journal*, 20 August 1816.

Chapter 2. In Search of a New Garden of Eden

> *Every person, however poor, may with moderate industry,*
> *become in a very short time a landholder; his substance increase from*
> *year to year; his barns are filled with abundant harvests; his cattle multiply*
> ....
> *Truly may it be said ... A paradise of pleasure is open'd in the wild.*
>
> —letter of Rep. Rufus Easton, 1816
>
> *All this I have seen, and 'tis too much; the delusion*
> *practiced upon the people of New England is beyond all human endurance.*
>
> — William H. Hand, *T'Other Side of Ohio* (1818)

Held back by frontier wars with the British and Native Americans, New England farmers were more inclined after 1816 to pack up their meager belongings and make the trek west. The "year without a summer" had shown that their long-term survival on land first tilled by their forebears two hundred years earlier was now in doubt.

Miraculously, almost no one died. With so many crops ruined, food was scarce, especially in northern New England, and many people had little to eat. But there was no massive famine. New Englanders came together to help each other through the long winter. Organizations like the Female Charitable Society of Concord, New Hampshire, handed out food to the neediest.[1] Countless fellow Americans responded generously to pleas for help such as this one from

---

[1] *History of Concord, New Hampshire, from the Original Grant in Seventeen Hundred and Twenty-Five to the Opening of the Twentieth Century*, ed. James O. Lyford (Concord NH: Rumford Press, 1903), 351.

Baltimore: "At a time when Divine Providence has denied to us the bounties of the earth, while Europe and America are both upon their knees and imploring bread, it surely becomes every man without distinction of sect or of party, to step forth with a generous ambition for the relief of his fellow men. Such awful visitations of Providence seem as if designed to inculcate the mutuality of our dependance [*sic*]." The crop failures brought much hardship, but also taught an important lesson about the parity of rich and poor in the eyes of God: "When the face of heaven frowns, and the earth in dreadful accordance parsimoniously withholds her bounties, the proudest monarch notwithstanding the blaze of his diadems, and the meanest slave are both upon a level; they are both beggars for daily bread."[1] This crisis, it seemed, conveyed both religious and social messages. First and foremost, many had neglected their obligations to God and had been punished for this failure. But why so many? What was their common sin?

Some preachers, like the Reverend Joseph Steward of Hartford, said it was not possible to understand why this barren year had occurred. Standing before a silent throng of his fellow Presbyterians on Connecticut's day of thanksgiving, 28 November 1816, he explained that mortal sinners could not comprehend God's motives; this punishment simply had to be foreborne as His "strange work," as much a part of His infinite love as the mercy He would show on other occasions.[2] This ordeal had reminded New Englanders of all classes of their lowliness before the Almighty. In His eyes, distinctions based on possessions and power meant nothing; even the rich and prominent could not evade divine judgment and had to meekly submit to God's will. This point, repeated in other pulpits around this time, obliquely criticized the hierarchies of social class, wealth, education, which dominated New England society.

But the most pressing lesson of the "year without a summer" was that good harvests could not any longer be taken for granted. Unusually cool summers the previous two years had already caused considerable hardship, and many families now stood on the brink of poverty. The shock of seeing late-summer corn tassels glisten with ice and cucumbers shrivel up for lack of rain had shaken their belief in nature's perennial renewal. The lives of hundreds of thousands of farmers and their families were now endangered. Another poor harvest like this one might be too much. Farming in New England might not be sustainable. They might have to move away.

As Samuel Griswold Goodrich, the self-taught son of a Congregational

---

[1] "By the Mails: The Call of Humanity," *Bangor Weekly Register*, 1 February 1817. This article was datelined "Baltimore, January 22."

[2] Rev. Joseph Steward, *A Sermon Delivered at the First Presbyterian Church in Hartford, on the State Thanksgiving, November 28, 1816* (Hartford: George Goodwin and Sons, 1817), 8.

minister raised in western Connecticut, would later write, the disastrous weather of 1816 caused a number of his friends and neighbors to lose "their judgment," convincing them that New England was turning into an Arctic zone, doomed to be frozen and inhabitable for centuries to come.[1] The "severe deprivation" they had suffered made them yearn for a less arduous and unpredictable life: "A ration of baked potato and the bloodstains from bare feet in the winter snow" were too much to endure any longer. Instead, they looked west — to Ohio, which they envisioned as a land of milk and honey where cattle grazed in pastures through the winter, only a few flakes fell, and spring came early. The desire to escape New England's brutal and inhospitable winters was "in itself a motive powerful enough to drive them on across the mountains into the wilderness." [2] As more and more left, a feeling of despair spread through towns and villages like an epidemic, and many hurriedly packed up their meager belongings and headed off to a better place.

Among those desperate to leave was the family of Joseph Smith. For several years they had eked out a living in the town of Sharon, Vermont, in the eastern part of that state. His father, also named Joseph, had run a dry-goods store and then concocted a scheme to get rich quick by buying up ginseng root from local farmers and exporting it to England, where it was valued as an aphrodisiac. But he had been cheated out of his profits from this venture and forced to sell his farm to pay off mounting debts. He and his wife Lucy became itinerant tenant farmers, moving seven times, from farm to farm around Sharon, during the fourteen years prior to the birth of their son, Joseph, in December 1805. His father taught school during the winters and farmed the rest of the year to support his eleven children. Hoping for better luck, the Smiths moved to Lebanon, New Hampshire, and then, in 1814, to the nearby Vermont town of Norwich. Their first year there the crops failed. Discouraged, all but one of the elder Joseph's six brothers moved on to New York, but he stubbornly vowed to try his luck on this land the next year, only to end up with another dismal harvest. He tried for a third time, in the spring of 1816, but the June snowfalls and August frosts destroyed his seedlings, and so he gave up. In the fall, the Smith family, including six of the children, relocated to a place in western New York with the historic name of Palmyra.[3] There, in a region already in the throes of religious revivalism, young Joseph

---

[1] Samuel Griswold Goodrich, *Reflections of a Life Time, or Men and Things I Have Seen*, vol. 2 (New York: Miller, Orton and Mulligan, 1856), 78.

[2] Harlan Hatcher, *The Western Reserve: The Story of New Connecticut in Ohio* (New York: Bobbs-Merrill, 1949), 71.

[3] Richard L. Bushman, *Joseph Smith: Rough Stone Rolling. A Cultural Biography of Mormonism's Founder* (New York: Knopf, 2005), 19-27. See also Stewart H. Holbrook, *The Yankee Exodus: An Account of Migration from New England* (New York: Macmillan, 1950), 48.

would find the Golden Plates on which he would build the Mormon religion.

The drought and unseasonably cold weather of 1816 had harmed farmers in several ways. First of all, it had greatly diminished their food supplies. They had grown little themselves, and the few staples for sale were too expensive, and so their ability to live off the land was now in question. Secondly, it had drained away most, if not all, of their modest incomes. They had no grain or seed corn to sell or barter. Farmers' other chief asset — livestock — depreciated as many animals were put up for sale, because their owners needed cash or did not have enough fodder to keep them alive during the winter. Thirdly, the extensive crop failure prevented farmers from paying down their debts — money borrowed to buy land. Regional indebtedness had been increasing for some time, as a result of several consecutive poor harvests. Even before the abysmal summer of 1816, farmers in upstate New York, for instance, were struggling to pay back what they owed because their fields were not producing very much. Noadiah Hubbard, a new arrival from Litchfield County, Connecticut, had complained that he was having trouble getting back the money he had lent to local farmers due to "very poor" wheat crop the year before, caused by hard frosts.[1] Farm debts were also mounting back in his corner of Connecticut. Many farmers were going out of business because credit was in such short supply, and they could not pay what they owed.[2] (Would-be independent farmers could not afford to buy any of remaining arable land and thus remained stuck at the bottom of the economic ladder.) In the town of Granby, on the Massachusetts border, frost had destroyed as much as three-quarters of the corn crop, and the long-lasting drought had dried up half the grasses on which cattle normally grazed and fed during the winter. Farmers' inability to earn much from their crops put creditors like young Oren Lee in a bind: his dream of opening up a small fabric factory, marrying, and starting a family had to be put on hold when a Granby farmer went bankrupt, leaving Lee with an unpaid debt of $400. "How shall I get along with it I know not," Lee wrote in his diary, "but I can hope some way will be provided."[3]

Indebtedness was not confined to New England. In Virginia, an aging Thomas Jefferson, perennially living beyond his means, could not pay his European creditors because of several adverse developments, culminating in what he bemoaned as the "most calamitous season for agriculture almost ever known." Many families living near Monticello could not afford bread

---

[1] Letter of Noadiah Hubbard to Julius Deming, 28 May 1816, Quincy Collection, Litchfield Historical Society.

[2] See, for example, letter of Charles and Frederick Deming to Julius Deming, 21 September 1816, Quincy Collection.

[3] Quoted in Mark Williams, *A Tempest in a Small Town: The Myth and Reality of Country Life — Granby, Connecticut, 1680-1940* (Granby CT: Salmon Brook Historical Society, 1996), 275.

because their fields had not generated enough cash for them to pay the soaring prices then being demanded — five times as much as usual. On top of this, the local wheat crop had been savaged by "such an inundation of Hessian fly as was never seen before." Like many other farmers, Jefferson had suffered from the cessation of trade with England during the War of 1812, leading to the collapse of numerous banks and a drastic dwindling of the money supply: credit was almost impossible to obtain. In Virginia alone, the amount of money in circulation had shrunk by half in just a year and a half. The collapse in prices for agricultural commodities had made it impossible to pay off some of his debts with what his fields produced, and he was now forced to consider selling off some "unprofitable" real estate instead. But here, too, he was stymied by the bad economy: a house and grounds in Richmond which had sold for over $6,000 just a short while before now could not fetch as much as $1,500. But Jefferson was relatively lucky. Those worse off than he had no way of paying their taxes or feeding their families.[1] A large number slipped into bankruptcy. The outlook for Virginia's rural residents was bleak.

And it was not only farmers who were hurting. Across the nation, skilled workers as well as shopkeepers and small manufacturers were also suffering financially because of three straight years of bad harvests. Disruption of commerce with Europe (and Canada) during the War of 1812 and more attractive economic opportunities further west, in the Ohio Valley, had cost them business. The shortages in 1816 came as the final blow to their already precarious livelihoods. Once the war had ended, England had resumed exporting goods, but at rock-bottom prices in order to stifle the growth of manufacturing in its former colonies. This strategy forced many fledgling factories and mills out of business. Shops across the region shut their doors, firing hundreds of workers. Among those affected were mechanics in the historic coastal Massachusetts community of Newburyport. Drawn there by openings at the shipyards, they had lost their homes in a devastating 1811 fire, which destroyed much of the downtown section.[2] Then, homeless and unemployed as a result of a sharp decline in the shipping business, Newburyport's mechanics formed a society in the winter of 1817 to encourage migration to the West, where their prospects seemed much brighter.[3]

Immediately after the "year without a summer," large numbers of Easterners would move to the frontier. This would mark the apogee of the

---

[1] Letter of Thomas Jefferson to Giovanni Carmigniani, 18 July 1816, *Works of Thomas Jefferson*, vol. 12, 33.
[2] Anonymous, *Particular Account of the Great Fire at Newburyport, May 31, 1811* (Newburyport MA: Shaw and Shoemaker, 1811).
[3] "From the *Newburyport Herald*: Emigration," *Christian Messenger* (Middlebury, VT), 29 January 1817.

first great westward migration in the nation's history. To understand how the alarmingly cold spring and summer of 1816 accelerated this exodus, one has to first comprehend this larger trend, which had been going on for several decades. Since the end of the Revolution — and the forging of peace treaties with Native American tribes in what is now the Upper Midwest — heading west had become safer and more appealing to Easterners. The chief attraction was cheap land. While Connecticut farmers had to dig deep in their pockets to come up with the fourteen-to-fifty dollars an acre that good local soil cost, acreage in western New York, Pennsylvania, and Ohio was selling for as little as three dollars.[1] Furthermore, it was of much higher quality than the shallow, largely depleted soils of New England.[2] Reports from travelers and migrants told of how effortlessly crops grew in the rich, alluvial lands along the Ohio, and how large a yield one could obtain with little labor. Typical of these seductive accounts was one written by New Yorker David Thomas, traversing the Midwest during the summer of 1816. Reaching Ohio in June, Thomas was surprised to find corn "nearly fit for the sickle," in growth a month ahead of what might he find back in the Finger Lakes. Wheat and rye were also more plentiful than he had ever seen before.[3]

New Englanders had been migrating to western New York and the Ohio Valley since the end of the Revolution, in search of better opportunities and more freedoms. This exodus had accelerated after Native American tribes and Western pioneers signed the Treaty of Greenville in 1795, granting whites the right to settle in Ohio. This territory had been hailed as a new "Garden of Eden." Leading the way there were some well-to-do men from Connecticut, acting as agents for land companies which had purchased government-owned tracts in northeastern Ohio — a three million-acre swath first called "New Connecticut," and later the Western Reserve. Most notably, the Connecticut Land Company bought up the eastern half of this large tract for $1.2 million. In 1796, it dispatched a surveying party, headed by Moses Cleaveland, a Yale-educated lawyer, to take possession of this newly purchased territory. At a point where the Cuyahoga River emptied into Lake Erie, these newcomers laid out the grid for what would become the Midwest's first industrial center.

---

[1] Richard J. Purcell, *Connecticut in Transition, 1775-1818* (Washington DC: American Historical Association, 1919), 150.

[2] One advocate of migration argued that it made more sense to move to virgin Western lands than to "resort to the laborious, expensive, and tardy process of replacing the energies of nature from the offals of the stall, and the stye [sic], the refuse of the kitchen and the barn . . ." John L. Tomlinson, "A Discourse on Agriculture," *Connecticut Journal*, 18 March 1818. Quoted in *The Peopling of New Connecticut: From the Land of Steady Habits to the Western Reserve*, ed. Richard Buel, Jr. (Hartford: Acorn Club, 2011), 98.

[3] David Thomas, *Travels through the Western Country in the Summer of 1816* (Auburn NY: David Rumsey, 1819), 102, 109, 100.

But it was its agricultural potential that made Ohio so desirable.[1] Enticed by ecstatic newspaper accounts and the hype of speculators, thousands of farmers and their families — mostly from areas where the soil was poor and unproductive — succumbed to what would later be called "Ohio fever" — the latest in a succession of such periodic "land-grab" crazes in American history.[2] By 1809, between fifteen-and-twenty thousand persons were living in "New Connecticut," most of them from the New England states. The coming of war in 1812 temporarily slowed the flow of settlers westward, but afterwards it was renewed, driven by a kind of "furor."[3] Once again, those hoping to turn a quick profit on their investments fueled this hunger for new land: "Business was done on a large scale, every inducement was offered, and words were not spared in the portrayal of rosy vistas for the prospective emigrant."[4]

Some early prospective settlers had come not to profit materially, but to Christianize the rugged hinterlands, where godly ways had been disavowed in the haste to clear and conquer the land.[5] Many of these zealots were new converts, carriers of the revivalist fervor then sweeping across New England.[6] Some were dissenters, chafing under the bit of a conservative Congregationalist hegemony.[7] But other migrants had no such higher mission; they were merely looking to escape stifling social constraints and financial obligations in staid, hidebound New England, and to "become

---

[1] Holbrook, *Yankee Exodus*, 26, 28.

[2] Successive waves of westward migration in the nineteenth century were commonly described as caused by "fevers." In addition to the one associated with the Midwest, there was the "Genesee Fever" of the mid-1790s, which induced thousands of New England farmers to relocate to western New York, as well as the "Oregon fever" of the 1840s and the California "gold fever" of that same period. Ralph J. Crandall, "New England's Migration Fever: The Expansion of America," *Ancestry* 18:4 (July-August 2000): 16. Use of this term suggests that many contemporaries viewed these mass migrations as a temporary madness, much like the *rage militaire* which impels young men to enlist in wars.

[3] Purcell, *Connecticut in Transition*, 140.

[4] Ibid., 145.

[5] Holbrook, *Yankee Exodus*, 29.

[6] Lewis D. Stillwell, "Migration from Vermont (1776-1860)," *Proceedings of the Vermont Historical Society* 5:2 (June 1937): 136. In some Vermont towns, as many as fifty persons a day would succumb to the "onslaughts of emotionalism" aroused by revivalist preachers. Some of these new believers decided to take the message of God to the frontier. This new religiosity was, in part, a response to the hardships Vermont and neighboring states were suffering: many embraced Christianity as a way to placate an "angry" God. T.D. Seymour Bassett, "The Rise of Cornelius Peter Van Ness, 1782-1826," *Vermont History* 10:1 (March 1942): 10. Van Ness — an attorney and future state legislator, chief justice of the Vermont Supreme Court, and governor — was one of many who joined the Congregational Church in 1817 and 1818.

[7] George W. Knepper, *Ohio and Its People* (Kent OH: Kent State University Press, 2003), 109.

somebody" on the fluid frontier.[1] These early migrants were lured by the chance to realize out west the unfulfilled promise of the American Revolution.[2] To members of the Yankee establishment like Timothy Dwight, the president of Yale, they were licentious, "shiftless, ne'er-do-well persons," a burden on his region's economy as well as a stain on its morals.[3] He was happy to see them go.

But their departure did not serve New England well. In several ways, it foreshadowed a long, irreversible decline. This was because power followed the settlers to the frontier. This was eminently so in the political arena. In the years after the Revolution, the Federalist Party, firmly entrenched in the New England states, had held sway in Washington. However, as new lands were settled and territories admitted to the Union, more seats in Congress and votes in the Electoral College were awarded to these new Western states. Because of the value they placed on individual liberty, equality, and freedom of opportunity, these states tended to eschew hierarchical Yankee values and vote for the upstart Democratic-Republicans, the party of Thomas Jefferson. This caused a national shift in the political alignment.

Prior to 1800, the Federalist Party had more than held its own by enjoying a comfortable majority of seats in the Senate, while control of the House had swung back and forth between the two parties. John Adams' defeat that year had been accompanied by a broad rejection of Federalist candidates in Congressional races as well: the party lost its majority in the Senate and was trounced in the House races, falling from sixty to thirty eight seats. The addition of two Western states — Kentucky and Tennessee — had helped to increase the Democratic-Republican base. Subsequently, two states had joined the Union — Ohio in 1803 and Louisiana in 1812. In the Buckeye State, both senators and all six members of the House apportioned to it in 1810 were Democratic-Republicans, and its growing population all but assured that additional members of that party would be sent to Washington. The situation was similar in Louisiana and prospectively so in all new states.

Western expansion would thus hasten the demise of the Federalist Party as a national force. Its New England bastion was not large enough to withstand this demographic shift. The best the Federalists could do was to

---

[1] Economic as well as political disadvantages to living in Connecticut persuaded many Democratic-Republicans to leave. See, for example, "For the Register," *Columbian Register*, 15 February 1817. The author of this article noted: 'The situation of a democrat in Connecticut has in it nothing inviting — No man, knowing what it is, would of choice subject himself to it."

[2] "For those who had it Ohio Fever was the excitement of a fresh start, of new communities to be founded, of new farms on rich soil, of new businesses to be organized, of money to be made, of churches to be built, of new friends who shared the same excitement of a new beginning." Fred J. Milligan, *Ohio's Founding Fathers* (Lincoln NE: iUniverse, 2003), xii.

[3] Quoted in Purcell, *Connecticut in Transition*, 139.

postpone the inevitable by discouraging people from leaving, as they had unsuccessfully attempted to block acquisition of the Louisiana Territory in 1803. When the Ohio Valley opened for settlement after the War of 1812, newspapers aligned with the Federalists waged an intense campaign against migration.[1] A sleuth of articles spelled out the hazards of life on the frontier — floods, disease, uncertain profits, the absence of civilization and morality.[2] These editorials scolded New Englanders who, suffering from a "deplorable species of madness," were setting out with no clear idea of where they were going, or how they would survive in the wilderness. They told of migrants who, "broken down in spirit and fortune" in some makeshift cabin, had longed for their "forsaken hearths" but died without ever seeing them again.[3] They lambasted "democrats" for "continually deceiving the people with flattering accounts of the western world . . . constantly slandering our institutions, by proclaiming the superiority of those of that country; they lead the ignorant to believe, against the evidence of their senses, that their own country is a barren rock, and that the fields of the West need only reapers to gather in the harvest . . ."[4]

For many Democratic-Republicans, the frontier was their salvation. Philosophically they preferred to live on equal terms with their neighbors. In

---

[1] Not all Federalists adopted this strategy, however. Some encouraged emigration through their "intolerance" to reduce the ranks of their Republican opponents and thus preserve their dominance of New England politics. See, for example, "To the Republicans of Connecticut," *Bridgeport Herald*, 3 April 1816. Quoted in *Peopling of New Connecticut*, 102. Efforts to persuade Republican families to leave the Nutmeg State intensified when the Federalists sought to take control of Connecticut's assembly and governor's office in 1817. See "Horrors of a Revolution," *Hartford Times*, 8 April 1817.

[2] The Federalist *Connecticut Courant* cautioned that the movement to the frontier was taking the country backwards, "from civilization to savageness." "The Brief Remarker," *Connecticut Courant*, 14 January 1817. Quoted in *Peopling of New Connecticut*, 114. A dearth of schools and churches in the Ohio Valley reinforced this perception. Shortly before his death, Timothy Dwight warned would-be emigrants about the "moral dangers" they would encounter outside the pale of Eastern civilization. Dwight, *An Address to the Emigrants from Connecticut, and from New England Generally, in the New Settlements in the United States*, (Hartford: Peter B. Gleason, 1817), 9.

[3] Purcell, *Connecticut in Transition*, 154. William Hand induced many New Englanders to reconsider migrating west with his *T'Other Side of Ohio*, published in 1818. In it, Hand recalled his visits with families who would invariably "fall into a strain of repining" over the decision they had made to leave the East, mixed with despair over not being able to return there again. Their disillusionment contrasted ironically with the optimism of the narrator's wife, who had proposed moving to the "pleasant land of Ohio, where are no north east winds and no winter; where spring and autumn are for ever [sic] blended, and summer and winter for ever banished . . . where corn grows 100 bushels to the acre if hoed, 50 if not hoed, and 25 if not planted." Hand, *T'Other Side of Ohio, or a Review of a Poem in Three Cantos* (Hartford: S.G. Goodrich, 1818), 34-8, 4-5.

[4] "From the *Worcester Gazette*: Emigration," *Hallowell Gazette*, 18 October 1815.

rural Western communities, the divisive factors of wealth, social status, and education would be ameliorated and — as Frederick Jackson Turner would spell out in his "frontier" thesis later in the century — democracy would flourish. Politically, the Democratic-Republican Party realized that the nation's geographical growth would increase its power. Hence, its leaders cheered on those in the New England and Mid-Atlantic states contemplating a move to the Ohio Valley. For their departure would translate into a "win-win" outcome— greater political power in the West as well as a weakening of the Federalist base, since every person who emigrated contributed to a decline in the population in these Eastern states — and, ultimately, of their representation in the federal government. Jefferson, for one, used the Lewis and Clark expedition to discover what the West had to offer and promote its settlement. Southern leaders, sensing the economic advantages of growing cotton and tobacco on lush Western soils, were among the most outspoken advocates of migration. In 1816, Andrew Jackson — basking in the glory of his victory over the British at New Orleans — declared that the nation's future well-being depended upon cultivating the rich, untrammeled lands of Tennessee, Alabama, Arkansas, and Mississippi. A year later, John C. Calhoun, then serving as Madison's Secretary of War, stated that Western migration was the way to "bind the Republic together" and urged settlers to "conquer space."[1] Northern Democratic-Republicans also relied heavily on newspapers to drum up interest in frontier states like Ohio. These publications played a large role in influencing public opinion since their number and circulation had mushroomed after the Revolution: even the smallest and most isolated villages were accessible. On the East coast, the two major parties marshaled the press to hammer home their positions on many issues, with Western settlement being one of the most important.[2]

A leading voice for migration was the *Eastern Argus*, published in Portland. It had been founded in 1803 — the year of the Louisiana Purchase — by prominent members of the Democratic-Republican Party in the District of Maine, for the express purpose of countering the "aristocratic" influence

---

[1] Mary Beth Norton, et al., *A People and A Nation: A History of the United States*, 9th ed. (Boston: Wadsworth, 2008), 244.

[2] Only a handful of the several hundred newspapers in those days could be considered politically independent. According to one survey conducted in 1810, there were 364 newspapers in the U.S., with 158 being classified as Republican, 157 as Federalist, and only forty nine "neutral." These figures from Isaiah Thomas, *History of Printing in America*, cited in Hezekiah Niles, *The Weekly Register*, vol. 1 (Baltimore: H. Niles, 1811), 116. A notable example of a politically "neutral" paper was the *Bangor Weekly Register*, which began publication in November 1815. Its editor, Peter Edes, announced that its intent was to publish pieces giving different perspectives on issues so readers could decide for themselves. See his "To the Public," *Bangor Weekly Register*, 25 November 1815.

of Federalists there.[1] Nathaniel Willis, a twenty-three-year-old Bostonian and son of a newspaperman, had been recruited to run this new venture and make its presence felt in this largely rural and independent-minded section of the country. Arriving with his young bride in tow, Willis pursued this goal assiduously. Starting in 1810, he regularly used the platform of his pages to tout the virtues of Ohio. In February of that year, the *Argus* ran a piece lauding the great changes which had taken place in that state over the past twenty years: a flood of settlers had transformed a "wilderness into a fruitful field." Farming had reached a "great height," and commerce and manufacturing were also booming. The *Argus* noted that Ohio now had a population of some two hundred thousand.[2] In the fall, the Portland newspaper printed a report from Chillicothe, Ohio, asserting that "Never, we believe, have crops been more abundant; and we could name one farmer near this place, who has raised upwards of nine thousand bushels of wheat!"[3] In the following years, the *Argus* kept its readers mindful of the "immense emigration" from the Eastern states, as well as the cheap land which was drawing more and more settlers like a magnet.[4] Western soil, it claimed, was a "rich mine, overflowing with wealth." Such glowing depictions became more persuasive after Maine farmers endured an unusually cold summer and poor harvest in 1815. These weather-related woes came on the heels of British military incursions into coastal towns and wartime disruption of trade. Many families found themselves in desperate straits.[5] Editorials in the *Argus* and other pro-migration papers played upon fears of a coming famine to convince them to leave for the frontier. Unemployed mechanics and other skilled laborers (such as those in Newburyport) were reminded that work was "plentiful" and wages high in burgeoning commercial centers like Cincinnati.[6]

With this relentless campaign, the *Eastern Argus* no doubt helped convince thousands of Downeast farmers to abandon their unproductive fields and head west.[7] (The number of persons leaving the District between

---

[1] Thomas N. Baker, *Sentiment and Celebrity: Nathaniel Parker Willis and the Trials of Literary Fame* (New York: Oxford University Press, 1999), 16.
[2] "State of Ohio," *Eastern Argus*, 1 February 1810.
[3] "Chilicothe [*sic*], Aug. 13," *Eastern Argus*, 27 September 1810.
[4] See, for example, "Emigration to Ohio," *Eastern Argus*, 12 December 1811, and Albert Gallatin, "Provision for the Ensuing Year," in the 19 December 1811 issue of this newspaper.
[5] William D. Williamson, *History of the State of Maine*, vol. 2 (Hallowell: Glazier, Masters and Co., 1823), 664.
[6] "From Liberty Hall, Published at Cincinnati, Ohio," *Eastern Argus*, 30 August 1815.
[7] John D. Post, *The Last Great Subsistence Crisis in the Western World* (Baltimore: Johns Hopkins University Press, 1977), 105. But Post also notes that many of these Mainers left in 1815.

1810 and 1820 has been estimated at ten to fifteen thousand.[1] This figure was equivalent to roughly six percent of Maine's population at that time.) But this editorializing also aroused the ire of Federalists. They fired back in kind. In January 1816, an editorial in the *Hallowell Gazette* accused the *Argus* of making "base attempts for party purposes to draw them [Maine farmers] into the ruinous rage for migration."[2] But Willis's paper was only one of many in the District and elsewhere in New England taking this position.[3] They liked to point out that Ohio and neighboring territories such as Kentucky were virtually free of Federalist influence: democratic-minded emigrants would be welcomed.[4]

Just how central politics was in making the case for emigration is evident in the speed with which Democratic-Republican papers abandoned their crusade once a serious proposal for Maine to secede from Massachusetts was advanced in 1815. With this goal on the horizon, their enthusiasm for leaving Maine evaporated, as the editorial focus shifted to building support for statehood.[5] Now papers like the *Eastern Argus* contended that moving west made no economic sense: farmers could do better by staying in Maine.[6] These Democratic-Republican publications echoed longstanding critics of resettlement such as the *Bangor Weekly Register*, which had insisted for years that Ohio was no land of milk and honey: farmers could make more profits there; the climate was unhealthy, and fever and "ague" all too common.[7]

But the larger truth here was that New England agriculture was in trouble: all the politically motivated arguments in favor of migration grew out of that reality. Recent crop failures — and the fear of more to come — underscored that the region *was* facing an existential crisis greater than any since the first

---

[1] Howard S. Russell, *A Long, Deep Furrow: Three Centuries of Farming in New England*, abridged ed., (Hanover, NH: University of New England Press, 1982), 148. Cf. Williamson, *State of Maine*, 666. Williamson notes that an unknown number of these émigrés eventually returned.

[2] Untitled article, *Hallowell Gazette*, 17 January 1816. This is one of the few articles from that period to invoke the phrase "Ohio fever."

[3] Crops like wheat and hemp grew in great abundance, it was reported. The latter was said to yield as much as 1,000 pounds an acre. See letter of Paul Hamilton to Thomas Newton, 21 January 1811, published in the *American Advocate* (Hallowell ME), 1 May 1811.

[4] See, for example, "Political: State of the Parties throughout the Union," *Rutland Herald* (Rutland VT), 14 March 1810.

[5] "Separation All Hail!" *Eastern Argus*, 21 August 1816. The convention actually took place in Brunswick, at the end of September.

[6] "The District of Maine — No. VI," *Eastern Argus*, 13 December 1815. By then Nathaniel Willis had left the newspaper and returned to Boston. See also "Miscellany," *American Advocate*, 3 February 1816. This cited a letter from a man in Cincinnati who regretted his decision to migrate there, as he could make bigger profits from growing potatoes back in Maine than corn in Ohio.

[7] For this Federalist perspective, see untitled article reprinted from the *Keene Sentinel* in the *Dartmouth Gazette*, 4 June 1817.

European settlers had arrived there in the seventeenth century. At root lay the undeniable fact that New England was not growing enough food to feed its population. It was equally true that New England's growing season was relatively short, that its farmland had been rocky and of mediocre quality since the days of the Pilgrims, and that now, after two hundred years of planting and harvesting, much of its topsoil had worn away. It was also true that the alluvial soils and climate in the Ohio Valley were superior to those in the Northeast. It was further true that thousands of New England farmers — or "yeomen," as they were commonly called — had fallen deeply into debt because of poor harvests and that they had little chance of getting out of this situation — or of staying out of jail if they couldn't pay back their loans.[1] It was true that roads in states like Connecticut were in such poor shape that trade outside the region was greatly hampered; there were no canals to carry goods to and from the interior either.[2] All these unpleasant facts were not lost upon New England's farmers and their families. Poor crop yields in 1814 and 1815 only made the precariousness of their existence more apparent.

These agricultural shortfalls were compounded by a tremendous gale which struck the coast of Connecticut with no warning on 22 September 1815, sending an eleven-foot high surge of water barreling up Narragansett Bay like a tsunami.[3] This "extraordinarily calamitous" storm smashed hundreds of homes and devastated the already foundering shipping business in southern New England. Its impact was felt all over the region, with millions of trees in New Hampshire knocked down, fruit orchards leveled, corn and wheat plants uprooted and swept out to sea, and cattle drowned by the onrushing torrent.[4] This event was so destructive that many New Englanders despaired of rebuilding their lives in its aftermath.

How many decided to migrate because of the *annus horribilis* of 1816 cannot be accurately determined because demographic data are collected only every ten years, through the federal census. Historians can only measure patterns over this relatively long period of time. Between 1810 and 1820, the population in many parts of New England increased only slightly, indicating there was a major outflow of residents — larger than at any other point in the region's history. Indeed, departures were so numerous that they nearly offset increases due to births and the arrival of newcomers — from Europe or other

---

[1] This policy of imprisoning indebted farmers was the cause of Shays' Rebellion, in western Massachusetts, during 1786 and 1787.

[2] A canal running across New York State to the Great Lakes had been proposed in 1807, and agreed upon in 1815, but construction of this waterway — the Erie Canal — would not begin until 1817. Rufus King, *Ohio: First Fruits of the Ordinance of 1787* (Boston: Houghton Mifflin, 1888), 347.

[3] New England did not experience a greater storm surge until the hurricane of 1938.

[4] Russell, *A Long, Deep Furrow*, 147.

states — during that decade. This spike in outmigration figured significantly in New England's loss of an estimated three-hundred-fifty thousand white males between 1800 and 1860.[1] While this massive relocation would greatly benefit the West and the nation's development, it would also hasten the end of New England's economic, political, and social preeminence. Fearing this consequence, governors such as Jonas Galusha of Vermont and Oliver Wolcott, Jr. of Connecticut tried to stem this "ruinous" Western migration.[2] They realized their states would face a bleak future if deprived of so many of their young men and women. [3]

The impact of outmigration in the 1810s was dramatic. Vermont offers a case in point. In the course of this decade, it added only 17,851 residents — a growth of less than eight percent. (By comparison, the population of the United States increased by one third during the same period.) Many of its towns and villages barely grew at all; some sixty of them shrank.[4] The town of Chelsea, first settled in 1784, had expanded its population by 275 percent between 1790 and 1800, and by nearly forty eight percent the following decade. But in the ten years after 1810, the rate fell to just over ten percent. (It rebounded to more than thirty three percent in the 1820s.)[5] In Tunbridge, where Joseph Smith and his family had lived before moving to New York, the population shot up by forty percent between 1800 and 1810, yet by 1820 the town had six hundred *fewer* residents than a decade earlier.[6] Other Vermont towns experienced a similar decline: Peacham's population fell from 1,301 to 1,294 and Shelburne's from 987 to 936. Several places, including Danby in the southern part of the state, were all but deserted.[7] In most locations, only minimal growth took place during the 1810s.[8] It would take the centrally

[1] Allan Kulikoff, "Migration and Cultural Diffusion in Early America, 1600-1860," *Historical Methods* 19 (Fall 1986): 164. This figure does not include boys under the age of 10.

[2] Barrows Mussey, "Yankee Chills, Ohio Fever," *New England Quarterly* 22:4 (December 1949): 449. As a candidate for governor in 1817, Oliver Wolcott, Jr., would proposed to slow this "desolating tide of Emigration" by imposing a five-year ban on emigration from the state. "Camillus," untitled article, *Columbian Register*, 8 March 1817. "The Times," *Times* (Hartford), 8 April 1817.

[3] At the start of the nineteenth century, two thirds of Vermonters were under the age of twenty six. Lewis Stilwell, "Migration from Vermont, 1776-1860," *Proceedings of the Vermont Historical Society* 5:2 (June 1937): 26.

[4] Post, *Subsistence Crisis*, 106.

[5] "Population Change in Chelsea and other Orange County Townships, 1790-1910," in Hal S. Barron, *Those Who Stayed Behind: Rural Society in Nineteenth-Century New England*, (Cambridge: Cambridge University Press, 1984), 27.

[6] Bushman, *Joseph Smith*, 27.

[7] Walter Hill Crockett, *Vermont: Green Mountain State*, vol. 3 (New York: Century History, 1922), 137. However, the northern and western parts of Vermont were the most affected by this outmigration. Stilwell, "Migration from Vermont," 137.

[8] Zadock Thompson, *History of Vermont, Natural, Civil and Statistical*. vol. 2, *Civil History of Vermont* (Burlington VT; Chauncey Goodrich, 1842), 209-10.

located Orange County over seven decades to make up for this depletion of its population.[1] These numbers tell a tale of economic hardship and despair. People were deserting Vermont because they could not foresee a secure future there.[2]

Western Massachusetts also saw a major exodus: towns like Sheffield, Egremont, Great Barrington, Stockbridge, Becket, Lenox, Pittsfield, and Lee added only a few residents during the 1810s, although their growth had already begun to slow a decade earlier. In the surrounding Berkshire County countryside, the population stagnated.[3] Neighboring Connecticut also lost tens of thousands of residents. Between 1810 and 1820, the state's population grew by only about thirteen thousand, or five percent. The number of people living in counties such as Litchfield remained essentially unchanged.[4] Many Connecticut emigrants headed for Ohio, establishing frontier outposts in the northern part of that state. For instance, Sandusky, near Lake Erie, was first surveyed in 1817 by a Danbury hat maker and land speculator named Zalman Wildman. He then convinced numerous neighbors and friends from Connecticut to purchase lots in the vicinity. In 1799, Captain John Leavitt and one of his cousins from Suffield, Connecticut, purchased some seventy eight thousand acres in the Western Reserve. His descendants became leading figures in state politics and government.[5]

Ohio's swelling population bears evidence to the impact of these New Englanders. Over this ten-year period, the state doubled in size — to over half a million. It is estimated that two thirds of this growth took place after the cold summer of 1816.[6] Although one cannot prove that the poor harvests that year caused this sharp upswing, the temporal correlation between these two events makes a causal relationship likely.[7] In Vermont, for example, the

---

[1] Bushman, *Joseph Smith*, 27.

[2] Oddly, despite the large scale of this relocation, it received almost no mention in the local press. In light of the equally sparse coverage given the disastrous harvest of 1816, it seems more likely that Vermont newspapers opted not to dwell upon bad news in order to lessen its consequences. Newspaper advertisements for land for sale told a different story, however.

[3] David D. Field and Chester Dewey, *A History of the County of Berkshire, Massachusetts, in Two Parts* (Pittsfield MA: S.W. Bush, 1829), 12.

[4] The county seat, also named Litchfield, had a population of 4,639 in 1810 but only 4,610 a decade later. Alain C. White, *History of Litchfield, 1720-1920* (Litchfield: Litchfield Historical Society, 1920), 206.

[5] Benjamin Lane, "Reminiscences of Benjamin Lane: Warren in 1799 — The First Settlers," *Western Reserve and Northern Ohio Historical Society* 30 (March 1876): 19.

[6] Ohio's population rose from 380,000 in 1817 to 581,434 in 1820. See James Oliver Horton, "Race and Religion: Ohio, America's Middle Ground," in *Ohio and the World, 1753-2053: Essays toward a New History of Ohio*, ed. Geoffrey Parker, Richard Sisson and William Russell Coil (Columbus: Ohio State University Press, 2005), 49.

[7] Demographic statistics show that the exodus from New England was

peak years of outmigration came directly after what some residents called the "Year of Two Winters."[1] Numerous anecdotal accounts of farm families passing through crossroads towns on their way west in 1817 attest to this mass relocation. For example, one resident of Easton, Pennsylvania, on the Delaware River, made a tally of the wagons he saw heading west in just one month in 1817, coming up with the figure of 511. Estimating that each carried a family of six, he concluded that 3,066 persons had crossed the river bound for the Ohio Valley in that period of time.[2]

Samuel Griswold Goodrich, a twenty-three-year-old Hartford bookseller and publisher, vividly described covered wagons leaving Connecticut that year. Each was piled high with nine or so small children, one at the mother's breast and "some crowded together under the cover, with kettles, gridirons, feather-beds, crockery, and the family Bible, Watts' *Psalms and Hymns*, and Webster's spelling-book — the *lares* and *penates* of the household." Many trudged mile after mile on foot, or rode in oxcarts. Others pushed handcarts mounded with their worldly goods. Goodrich could see grim determination in their faces — not expressions of hope and joy, but of desperation. Many were so hungry and destitute that they reached out their hands to beg for food or money.[3] According to one newspaper, the flood of such despondent families, early in 1817, was ."constant" and from "every part of the state."[4] During the months of September and October, five hundred persons a week — mostly Yankees — passed through Albany on their way west.[5] From June 1st to early November, an observer in Brattleboro, Vermont, reported that 376 adults and 240 children had passed through this town, accompanied by thirty-six ox teams and forty-two horses. He was ascertained that the bulk of these itinerants came from Vermont, Massachusetts, and New Hampshire.[6] Another thousand went through Springfield, Massachusetts, that summer — 140 during one week in September alone.[7] Among those Yankees who had

---

greatest where the temperatures were the coldest. In Massachusetts and Connecticut, the biggest population losses were experienced in largely upland counties of Berkshire and Litchfield, respectively. See Joseph B. Hoyt, "The Cold Summer of 1816," *Annals of the Association of American Geographers* 48:2 (June 1958): 125-6.

[1] Stillwell, "Migration," 134-5. Another important influence was President Madison's message to Congress in December, revealing that the United States was now at peace with Native American tribes living within its borders. R. Douglas Hurt, *The Ohio Frontier: Crucible of the Old Northwest, 1720-1830* (Bloomington: Indiana University Press, 1996), 362.

[2] Mussey, Yankee Chills," 450.

[3] Goodrich, *Recollections*, 79.

[4] "For the Register," *Columbian Register* (New Haven), 15 February 1817.

[5] "Western Migration," *Vermont Republican* (Windsor VT), 22 December 1817.

[6] Mussey, "Yankee Chills," 449.

[7] Ibid., 450. See also "Emigration," *New York Post Herald*, 7 October 1817. This article cites the *Springfield Federalist* as its source.

caught "Ohio fever" was Rutherford Hayes, father of the eponymous future president. A storekeeper in Brattleboro, he ended up in the central town of Delaware, where his illustrious son was born. On Independence Day, Cincinnati saw more persons arrive than ever before in its brief history.[1] The new arrivals kept coming by boat all summer long and even into the winter, borne down a rain-swollen Ohio River.[2] According to newspapers, "Ohio fever" was continuing to "rage."[3]

Indiana, a recent addition to the Union, was also drawing more settlers. A western New York farmer joined a party of Mainers on the Allegheny River — a "cavalcade" of twenty wagons and 116 persons from the same town, all bound for the brand-new state.[4] A Massachusetts visitor to Vincennes wrote of having encountered two parties of settlers, totaling 150, who had taken up residence outside this former French fur trading post. The local land office was handling some fifty requests a day, at prices as high as thirty dollars an acre, and the speculators in town were so numerous there were not enough hotel rooms to house them. All were eager to grab a piece of what one enthusiastic Easterner claimed was the "finest agricultural situation in our western world."[5] The territory of Illinois also became a major destination once the federal government announced that lands there would be awarded as bounties for military veterans, starting in October 1817.[6] Businessmen saw potential in this new land and quickly grabbed it up. Among those responding to Illinois' siren call were the four Collins brothers — Anson, William, Augustus, and Michael — all from Litchfield. They came to a place near St. Louis called Unionsville, having bought up the holdings of its original settler, and soon had erected a saw mill, flour mill, distillery, and store there. In a few years, the enterprising brothers scattered across the state and figured prominently in growing its economy.[7]

---

[1] "Singular Arrival," *Newburyport Herald*, 1 August 1817.

[2] Untitled article, *Providence Patriot*, 6 December 1817. This reported that the flooding Ohio River had brought a "vast crowd of strangers" into Cincinnati in December.

[3] See, for example, untitled article reprinted from *Keene Sentinel*, *Dartmouth Gazette*, 4 June 1817. This piece was also published by several other New England newspapers not supportive of outmigration. Some, like the *Bangor Weekly Register*, were openly skeptical about the wisdom of moving west and urged families expecting a "land of milk and money" to think twice.

[4] "Genessee Farmer," untitled article, *Vermont Republican*, 22 December 1817. Indiana was admitted to the Union in December 1816.

[5] "Rising Importance of the Western States," *Rhode-Island American* (Providence), 29 November 1816.

[6] The General Land Office made this announcement in September. See untitled article, *Register* (Windham CT), 1 January 1818.

[7] Holbrook, *Yankee Exodus*, 65. See also http://penelope.uchicago.edu/Thayer/E/Gazetteer/Places/America/United_States/Illinois/_Texts/DRUOIH/Southern_Illinois/9*.html. Influenced by their former neighbor, The Reverend Lyman Beecher, the Collins brothers would subsequently

All told, some fifteen thousand Yankees were thought to have left for the frontier in 1817. An equal number came from the Mid-Atlantic states, joined by an estimated ten thousand foreigners, mostly British farmers.[1] A year later, according to a spokesman for the Philadelphia Society for Promoting Agriculture, the "torrent of emigration" was still growing, having already produced "disastrous and fatal" consequences. Entire townships were being depopulated.[2] Southerners were also on the move. Farmers in western North Carolina were reported to have walked away from their lands and headed west into the mountains because of the crop failure the previous summer. [3] Further inland, large contingents from Kentucky and Tennessee — as many as twenty thousand men, women, and children — had begun the trek toward richer, more abundant farmland. [4]

Caravans of ill-fed, ill-clad, and ill-tempered Yankees deepened the ruts in trails across New York and Pennsylvania through the summer of 1817 and into the fall, even as an ample harvest appeared to negate the compelling reasons for leaving, and many of their erstwhile neighbors, having failed to find any "promised land," were returning home. The magnetic pull of the Western "Eden" was so strong it overcame inertia, natural reluctance to start a new life, the anticipated rigors of the long journey west, and the risks of failure once one arrived.

It seemed as if a compulsion was driving them — a welling fear that, if they stayed behind, they would be lost; they would miss out on this one chance to better their lives. Individual doubts were swept aside, and the pressure to join this mass migration became irresistible, like that pushing

---

spread the temperance movement to Illinois, first destroying their distillery and then barring the sale of liquor on properties they sold.

[1] Many of them had been inspired to move to Illinois by an enthusiastic Quaker émigré from Surrey named Morris Birkbeck. His laudatory *Notes on a Journey in America from the Coast of Virginia to the Territory of Illinois* (1817) went through eleven editions in the three years after its initial publication. Many English immigrants headed for the settlement he had established at Albion, Illinois. For evidence of Birkbeck's impact, see "Spirit of Emigration," *Newport Mercury* (Newport RI), 10 October 1818. This article, quoting the *Pittsburgh Gazette* of 29 September 1818 noted: "Our shores exhibit one of the most animated scenes of bustling emigration we have ever witnessed — the beach of the Monongahela has been for several days completely lined with flat boats, destined for the Illinois and other districts below the Falls of Ohio." Pointing out that most newcomers are English, the article concludes that the "rage of emigration" from that country appeared to "pervade the whole kingdom . . . We suspect we are indebted to Mr. Birkbeck's flattering representation for this increase."

[2] "Address to the Citizens of Pennsylvania on the Importance of a more Liberal Encouragement of Agriculture." Quoted in Steven Stoll, *Larding the Lean Earth: Soil and Society in Nineteenth-Century America* (New York: Wang and Hill, 2002), 76.

[3] C. Edward Skeen, "The Year Without a Summer': A Historical View," *Journal of the Early Republic* 1:1 (Spring 1981): 65.

[4] "Western Migration," *Vermont Republican*, 22 December 1817.

spectators out of a burning building. Contemporary observers like Samuel Goodrich aptly characterized this exodus as a "sort of stampede," "torrent," "panic," or "fever."[1] The sense of urgency felt by so many supports the thesis that the specter of poverty and starvation following the "year without a summer" was propelling it.[2]

Finally, by the fall of 1817, migration "fever" began to ease, as cooler heads reassessed the pros and cons of abandoning New England for an uncertain Western future. Newspapers welcomed this return to "sanity." The "wild spirit" of the spring, which was "almost epidemical," had subsided. In retrospect, it seemed the "height of madness" to have fallen for the speculators' "most deceptive" sales pitch.[3] As early as July, the *Dedham Gazette* credited the "fruitfulness" of the growing season in New Hampshire and neighboring states with having checked the "fever of emigration."[4] In an article which was widely reprinted, the *New Hampshire Patriot* mocked those who had gone west thinking they would find an easy living there and had ended up living as "vagabonds."[5] The Federalist *Hallowell Gazette* reported on two families who had left Kennebec County (in the southern part of Maine) the previous September, egged on by the "false representations" of speculators, and arrived in western New York with only a few coins in their pockets. Reduced to a "beggarly situation," they had had to turn tail and go back to Maine.[6]

Memoirs and diaries describing this historic migration contain many such stories of disillusionment, suggesting it was not merely a figment of editorial imaginations. In June 1816 — just at the start of the chilly summer — a twenty-six-year-old aspiring lawyer by the name of Nathaniel Dike had set out from Haverhill, Massachusetts, with his heart set on Ohio— only to encounter misery and poverty along the way. Furthermore, he did not much care for the people he met en route, especially the Pennsylvania Dutch farmers who, aside from being ill-mannered, struck him as "the most ignorant, bigoted, surly, filthy beings on earth." Despite his dyspeptic disposition, Dike continued on his way and reached Ohio by mid-July. However, his mood was

---

[1] Goodrich, *Recollections*, 79. See also Russell, *Long, Deep Furrow*, 148, and "Emigration," *Dedham Gazette* (Dedham NH), 25 July 1817. The latter article asserted that 1816 had induced a "panic" among the young, who believed that only "immediate flight can save them from immediate death."
[2] Indications that a shortage — and resulting high cost — of wood fuel forced many poor families to emigrate support this thesis. See, for example, "Observer," *Connecticut Courant*, 31 December 1816.
[3] "Emigration," reprinted from the *Dedham* Gazette in the *Salem Gazette* (Salem MA), 16 September 1817.
[4] "Emigration," *Dedham Gazette*, 25 July 1817.
[5] "Emigration to the West," reprinted in the *American Advocate*, 16 August 1817.
[6] "Tide Turning," *Hallowell Gazette*, 15 October 1817.

not improved by what he found there. His fellow settlers were of "humble origin," little educated, barely better than brutes, and "immensely inferior" to the people among whom he had grown up. Economically, Ohio was no better off than New England, he wrote to his brother. ("The groans of poverty are frequent & audible, and still thickening.") His verdict seem to be borne out by the fact that many new arrivals, unable to pay the sky-high land prices now being commanded, were moving on to Indiana, Illinois, and Missouri.[1]

Far from home, on alien soil, their futures far from bright, many uprooted Yankees like Dike (who had broken down and cried over his predicament) felt a poignant longing for their home states. Some expressed this sentiment in verse, as did the anonymous author of these lines, published in a Connecticut paper, perhaps to give would-be migrants reason to reconsider:

> What land is that, where onions grow;
> Where maidens' necks are white as snow,
> And cheeks like roses red you know;
> Where jonny cakes are bak'd from dough;
> That land where milk and honey flow?
> Connecticut.[2]

While parties of settlers continued to leave during the fall, the anxiety which had prompted the mass departure earlier in 1817 had dissipated. The frontier still beckoned, but it was the call of opportunity rather than the wail of necessity which now set the wagon wheels in motion. Those who had impetuously departed, without a clear plan for the future, were now considered foolhardy. Reporting on the unusually large number of persons passing through Springfield, Massachusetts, in September, a reporter for the *Connecticut Courant* berated these "honest yeomen" for being tricked by speculators and predicted their "deplorable species of madness" and rash "infatuation must result in the wretchedness and misery of multitudes." That disappointment and regret awaited them out west "was certain as the fixed course of nature."[3] But the *Courant* was hardly an unbiased party. For decades it had served as the editorial voice of Connecticut's Federalist Party and had overlooked no opportunity to bemoan the departure of people and wealth from the Nutmeg State.

Without a doubt, Ohio and neighboring states had not proven to be a "land of milk and honey," as the proselytizers for migration wanted New Englanders to believe. Like anywhere else, the Ohio Valley suffered from the vagaries of climate, land prices, business cycles, natural disasters, disease,

---

[1] Letters of Nathaniel Dike to John Dike, 21 June 1816, 15 July 1816, and February 1818, "Nine Letters of Nathaniel Dike on the Western Country, 1816-1818," *Ohio History* 67:3 (July 1958): 196, 203, 208, 215, 216.

[2] Untitled, unsigned poem, *Register* (Windham CT), 1 January 1818.

[3] Untitled article, datelined Springfield, 25 September 1817, in the *Gazette* (Portland), 14 October 1817.

and war. The state's money supply shrank drastically after the War of 1812, and the economy nearly collapsed after the federal government then began requiring prospective landowners to pay in cash.[1]

Despite all these disappointments out West, the fact remained that its lands *were* more productive and affordable than New England's. In the coming decades, the future of American agriculture lay in the Ohio Valley. If nothing else, the "year without a summer" had made this clear. States like Connecticut might try to improve their agricultural yields — by planting other crops and using better growing techniques — but these were rearguard actions, only able to keep their economies going until a transition into some new form of production could be developed. As will be discussed in a later chapter, the decline of farming in New England, accelerated by the poor harvests of 1816, would pave the way for the region's manufacturing sector to grow exponentially and flourish. During the same time, the states of the Old Northwest became America's primary breadbasket. By 1860, Ohio would be second in the nation (after New York) in number of farms — with over one hundred seventy-three thousand producing staples for local and national markets. Illinois and Indiana would rank fourth and fifth (after Pennsylvania).

The rise of Ohio, Illinois, and Indiana as food-producing and -marketing states was spurred by the arrival of thousands of ambitious, experienced farmers from the East coast, but also by the improvement of transportation links between the interior and cities in New York, Pennsylvania, and New Jersey, as well as in New England. Without a doubt, the single most important development was construction of the Erie Canal. Talked about for more than a decade, the canal became a reality largely through the efforts of newly-reelected New York governor DeWitt Clinton, who convinced his state's legislators in 1817 to allocate the staggering sum of seven million dollars (forty six billion in 2014 dollars) to construct this 363-mile-long waterway—the first public works project in the United States. [2] At that time, western New York was largely an uncharted wilderness, without many roads or inhabitants.[3] While many scoffed at "Clinton's Ditch" and predicted it would never be completed in their lifetime, the canal, in fact, opened eight years later. Almost immediately it exceeded the dreams of its backers: tremendous quantities of Eastern manufactured goods flowed

---

[1] The conflict with England had left Ohio deeply in debt and prevented the state from investing in badly needed infrastructure — including the roads and canal which were required to accelerate the flow of migrants.

[2] The state senate in Ohio declined to help pay for the canal, with some members arguing that its construction was properly a federal matter. Hurt, *Ohio Frontier*, 389.

[3] The six counties lying between Pennsylvania and Lake Ontario had a total population of only some twenty three thousand in 1810.

inland while barges passing in the other direction brought foodstuffs to New York and states up and down the East coast. Thanks to this new water link, the population of Ohio doubled during the first five years of its existence. A dramatic drop in freight rates (from one hundred dollars a ton to ten dollars) proved a boon to farmers in the Ohio Valley. Whereas a mere four thousand tons of wheat was shipped across New York State in 1829, that figure had jumped to one million tons within eleven years.[1] Having so affordable access to Eastern markets greatly boosted commercial agriculture in states like Ohio, Indiana, Illinois, and Michigan. In addition, the opening of the Erie Canal enabled Eastern manufacturers to ship their goods profitably to this rapidly growing region of the country. This ability, in turn, encouraged the construction of factories and mills along the Atlantic seaboard and eased New England's evolution from an agricultural-based economy to a mixed one. The existence of this east-west waterway assured that the people of New England would never again have to depend for their survival solely on what their farms produced.

Large-scale settlement of the Ohio Valley was a direct consequence of New England's crop failures in 1816. The relocation of so many Americans to the interior had as great an impact on the growth of the United States as the doctrine of Manifest Destiny would have later in the century, when pioneers crossed across the Great Plains and the Rocky Mountains to reach another "promised land" on the Pacific. While this earlier migration figured so prominently in economic terms, it also helped to shift political power away from the Federalist Party into the hands of more democratic forces.

This occurred on both the national and regional levels. In New England, the Federalists' inaction during this agricultural crisis— their stubborn adherence to the "steady habits" of their forebears — helped turn voters against them. Their paternalistic manner of governance, based on trust in God's benevolence, no longer seemed adequate to deal with the harsh new realities confronting the region. Instead, New Englanders now sensed they had to assume responsibility for their own well-being. A spirit of independent thinking and rational inquiry took hold and dislodged the aristocratic and theocratic elites which had ironically so long dominated the "cradle of American democracy."

---

[1] "Construction of the Erie Canal: DeWitt Clinton," *Dutch Entrepreneurs*, 6. http://www.nnorg/nni/Publications/Dutch-American/dabook/Chapter%206.pdf.

Chapter 3. A Real American Revolution

> *Let a nation assume the purest republicanism, and work into their consti-*
> *tution the most refined principles of liberty, and then discard the doctrines*
> *and crush the institutions of religion, and their fine wrought threads will be*
> *wiped away, like a cobweb, and chains will supply the place.*

— Pastor Pliny Dickinson, June 1816

> *. . . rulers are no longer regarded, as the viceregents [sic] of heaven, with au-*
> *thority to execute their own will and pleasure, on earth, but have dwindled*
> *down to mere servants of the people, accountable to them for their conduct*
> *— the blind reverence for ancient institutions is abolished . . . The darkness*
> *is rapidly rolling away.*

— Josiah Quincy III, September 1816

If New England's weather was prone to great volatility, the same could scarcely be said about its political climate. Since the founding of the Republic one party — the Federalists — had ruled with a regal sense of entitlement, based on a Calvinist notion of predestination. But loyalty to these men of higher status hinged upon continuing stability and trust in their judgment. The crop failures of 1816 revealed their fallibility and breathed new life into democratic forces in the region.

On Election Day in Connecticut — the only state not to alter its constitution since the colonial era had ended — ministers dressed in black sallied forth from their New Haven meetinghouse looking like a flock of penguins about to negotiate an ice floe. Jauntily they strode through the center of town, puffing on their long ivory pipes and sipping tankards of ale.[1] Buoyed by these spirits and the festive mood, they would discuss the men they considered good candidates for governor, the upper house of the legislature, and other state offices. These were men they knew well, and so the selection process was casual and jovial. Indeed, many of the chosen ones were close relatives. When Roger Griswold became governor of Connecticut, in 1811, he was following in the footsteps of his maternal grandfather, his father, and his uncle. His cousin, Oliver Wolcott, Jr., would assume the same office a few years later. Usually there was no real contest because the incumbent Federalists were running unopposed. Once in office, they stayed there until they ran for another one, or died.[2] Then the lieutenant governor — the heir apparent — took over, as a prince replaces a king.

In this "land of steady habits," royalist rituals persisted.[3] This remained so even after power was finally relinquished by Connecticut's Federalists in 1817, and the younger Wolcott became governor. On that day — May eighth — members of the defeated party paraded through Hartford. Some bore symbols of the state's colonial past, to remind cheering crowds of the unbreakable bond between English power and Puritan faith. The marchers, including military guards in British uniforms, had lustily sung "God Save the King" so many times before that its outmoded lyrics no longer aroused any "sense of impropriety." The heads of the state's Congregationalist establishment who spoke on this day wasted no breath denouncing the victorious, renegade "Toleration" party and its democratic principles, but, instead, fondly recalled a form of government "diametrically opposed" to this upstart one. They were resolute in their desire — as one hostile Democratic-Republican newspaper put it — to "destroy the free spirit of our people, and to keep alive that ridiculous veneration for the authority of Kings, and an established hierarchy, which have so long disgraced the human reason." One sermon that day, noted the same paper, was more appropriate for Roman Catholics in imperial Spain than for a "metropolis of this enlightened state." The notion that governments were not built upon a social compact among their citizens, but upon "the ordinances of God," and that the people ought to

---

[1] New Haven and Hartford were co-capitals of Connecticut for most of the nineteenth century.

[2] Jonathan Trumbull, Sr., the state's first governor, served for fourteen terms before retiring in 1784. He died the following year, aged seventy-four. His successor, Matthew Griswold, was defeated after one term in office. Samuel Huntington died in office in 1796 during his tenth term. Oliver Wolcott, Sr., died in office after serving one year. Jonathan Trumbull, Jr., was elected twelve times before passing away in 1809. John Treadwell was defeated in his bid for re-election in 1811.

[3] Shortly after the end of the War of 1812, members of Hartford's ruling elite turned out to welcome England's Vice Admiral Henry Hotham and his commanders who had recently been blockading the port of New London. Sean Wilentz, *The Rise of American Democracy: Jefferson to Lincoln* (New York: Norton, 2005), 183.

bow before kings and other rulers with "reverential awe," struck the writer as too "absurd and dangerous to be regarded with indifference."[1]

But this was how Connecticut — and, to a lesser degree, Massachusetts — operated in those days.[2] Deference to the powers-that-be had a longer history than allegiance to the infant American republic and its "Blessings of Liberty." It derived from a centuries-old belief that the world was governed by immutable, God-given laws, which controlled everything from the fate of armies to the fluctuations of the barometer. This respect for authority was reinforced by a confluence of state and ecclesiastical power. Both the Federalists and the Congregationalists espoused the same hierarchical world view: they honored tradition; disparaged those who challenged their hegemony; believed that only a select few (men of good Puritan stock) possessed the wisdom and judgment necessary to lead; and rejected change as inimical to a stability which they felt had served the Nutmeg State well for generations. These two power structures were upheld by a combination of custom and law. Taking advantage of a suffrage limited to white male property owners, the Federalist Party routinely won virtually all state and local elections.[3] Out of respect for the state's Puritan founders, the Congregational Church remained the nation's only legally established church, with taxpayers required to pay its ministers regardless of their own religious beliefs. Although Congregationalists and Federalists theoretically held sway over different spheres of life, they actually ruled as one, because, in Calvinist America, any distinction between church and state was not only wrong, but illogical. For God oversaw both. Highly able men ruled over their fellows because the deity had given them the brains and talents to do so, that they might interpret and carry out His will.

To quarrel with this civil order was to question the wisdom of the Almighty, and one courted divine punishment and eternal damnation by doing this. Everything in the universe occupied its proper place in the "Great Chain of Being," as God had intended. This order sustained His

---

[1] Untitled article, *Times* (Hartford), 13 May 1817.

[2] Early in the nineteenth century, Federalist candidates for office were chosen by county representatives of the party. These were appointed by a central committee made up of seven prominent (and unelected) Bostonians. The existence of this committee was kept secret as its modus operandi clashed with New England's tradition of open government through town meetings. Harrison Gray Otis, *Otis' Letters in Defense of the Hartford Convention and the People of Massachusetts* (Boston: Simon Gardner, 1824), 289-90.

[3] In statewide elections held between 1800 and 1816, Connecticut's Democratic-Republicans never gained more than forty three percent of the votes and usually attracted a much smaller following. Andrew Siegel, "'Steady Habits' Under Siege: The Defense of Federalism in Jeffersonian Connecticut," in *Federalists Reconsidered*, ed. Doron Ben-Atar and Barbara B. Oberg (Charlottesville: University Press of Virginia, 1998), 202.

plan; disorder was a kind of sin.[1] In Congregationalist services, for example, families sat in the same pews for generations, affirming the heredity aspect of "natural aristocracy."[2] Geographically removed from other states and almost entirely of British Protestant heritage,[3] New England hewed to these values even after they were discredited elsewhere. Hence, democracy was slow to establish a foothold there.[4] It would take a major blow to the Yankee ethos for that to happen.

Belief that their lives served a higher purpose — and had to be lived in accordance with it — had brought the first boatload of Puritans to the New World in 1620. It had led their leaders to take up a quill pen and scribble their signatures on a piece of parchment later known as the "Mayflower Compact." Reaffirming their mission to found an outpost on this side of the Atlantic "for the Glory of God," these colonists agreed to "combine ourselves together into a civil body politic; for our better ordering, and preservation and furtherance of the ends aforesaid; and by virtue hereof to enact, constitute, and frame, such just and equal laws, ordinances, acts, constitutions, and offices, from time to time, as shall be thought most meet and convenient for the general good of the colony . ." To achieve these goals, these religious dissidents promised "all due submission and obedience" to their designated leader.[5] A decade later, the colony's newly elected governor, John Winthrop, reiterated this credo in a sermon entitled "A Model of Christian Charity." Here he asked that all his followers "be knitt together in this worke as one man," that they " rejoyce together, mourne together, labour, and suffer together, allwayes haveing before our eyes . . . our Community as members of the same body," united by their love for God .[6]

Only by obeying God's word and "cleaving to him" could the colonists hope to survive in this New World. Winthrop expected his own authority to be equally honored. He stood before the others as the chosen leader just as the Puritans saw themselves as God's chosen people. Realizing that bickering among the colonists would lead to ruin, Winthrop — a wealthy land owner — rejected proposals to extend the suffrage and other rights to members of

---

[1] Richard L. Bushman, *From Puritan to Yankee: Character and the Social Order in Connecticut, 1690-1765* (Cambridge: Harvard University Press, 1967), 4-13.
[2] James C. Welling, *Connecticut Federalism, or Aristocratic Politics in a Social Democracy* (New York: New York Historical Society, 1890), 38.
[3] In terms of ethnic origin, Connecticut was then the "most homogeneous" state in the Union. Siegel, "'Steady Habits,'" 221.
[4] Anson Ely Morse, *The Federalist Party in Massachusetts to the Year 1800* (Princeton: University Library, 1909), 9.
[5] The compact was deemed necessary after some of the Puritans on board the *Mayflower* had argued they were not bound by their original commitment once the plan to settle in Virginia had been scrapped in lieu of founding a colony in Massachusetts.
[6] This version of Winthrop's sermon is available at http://www.mtholyoke. edu/acad/intrel/winthrop.htm.

the Massachusetts Bay Colony, declaring them unworthy of such privileges. He likewise objected to any laws limiting the power of magistrates to rule as they deemed fit. Winthrop also opposed any suggestion of democracy, which he considered heresy and the "meanest and worst of all forms of government."[1]

Over the intervening century-and-a-half Massachusetts had moved away from these theocratic roots. Under the British crown, competence and profits had come to matter more than fealty to God. When Massachusetts became a royal colony, in 1691, the proviso that only members of the Congregational Church could vote was rescinded. The first governor of this expanded province was a ship's carpenter named William Phips — born in Kennebec, Maine, the youngest of twenty six children. He came to this position not because of his family's blood line but because he had salvaged loot from ships sunk off the coast of Haiti. By the 1760s, Massachusetts' royal governor was secretly planning to give the colonists a voice in the British Parliament — a shocking proposal in those days. But the colony's prospering commercial class did agree it ought to have a voice in its affairs. The Crown's insistence on taxing these merchants without their approval turned them into revolutionaries. Once the Revolution was over, however, the ruling class in Massachusetts grew concerned about the excesses of democracy — not only in France, but also in its own backyard. In the summer of 1786, impoverished farmers and veterans in the western half of the state had taken up arms to protest debtor laws, in the first open challenge to state authority in American history. This insurrection had frightened Boston's Brahmins as much as the sight of a column of Redcoats on Bunker Hill.[2]

In the ensuing years Massachusetts voters backed candidates who promised to keep a steady hand on the state's tiller. They upheld "a wholesome regard and respect for those whose tradition it had been to be the leaders and guides of their community, whether ministers or gentry."[3] Unlike Connecticut, this brand of conservatism did not depend upon an alliance between the Congregational establishment and the Federalist Party. Since the days of John Winthrop and Cotton Mather, succeeding generations had quietly disavowed the Puritans' gloom-and-doom deity. The coming of Universalism had freed believers from dependence upon clergy to interpret the Bible and encouraged a degree of free thinking which would have made Winthrop apoplectic.[4] A major shift away from the Calvinist

---

[1] Quoted in Samuel Morison, *Builders of the Bay Colony* (Boston: Northeastern University Press, 1930), 92.

[2] Morse, *Federalist Party*, 44, 50.

[3] Ibid., 69. Yet, between 1803 and 1813, Massachusetts voters regularly elected Democratic-Republican candidates to the House of Representatives in nine of the state's seventeen Congressional districts.

[4] Nathan O. Hatch, *The Democratization of American Christianity* (New Haven:

outlook occurred when the trustees of Harvard— whose presidents since 1635 had almost all been orthodox Puritan clergymen — chose a liberal-minded Unitarian named John Thornton Kirkland to run the college. This set a precedent which would be honored for the next 123 years, effectively cutting the ties between Puritanism and education at America's leading center of learning. Henceforth, scholarship at Harvard would pursue an independent course — often in opposition to Church teachings. Kirkland established professional schools (of law and divinity), added programs in the natural sciences, doubled the size of its library, attracted students from all over the country, and elevated Harvard to the academic prominence it has maintained to this day.

By contrast, in Connecticut the unifying of state, religious, and educational power constituted an indomitable monolith. Perhaps no single figure better embodied this unity than the Reverend Timothy Dwight. Born in Northampton, Massachusetts, in 1752, Dwight was the son of a prominent merchant, grandson (on his mother's side) of the fiery Calvinist preacher Jonathan Edwards, and great-great-grandson of the early Puritan leader Thomas Hooker. A precocious and devout youngster, he had fulfilled all the requirements to enter Yale by the time he was eight, but had to wait another five years before the college would accept him. His freshman year, in 1765, a pious Dwight was appalled to find it addled by godless behavior: well-heeled students entertained in their rooms, drank, neglected their courses, and gambled through the night. Dwight managed to resist most of these temptations by focusing on his studies, rising each day before dawn to translate a hundred lines of Homer by candlelight. He graduated at the top of his class. During the War of Independence Dwight served briefly as an army chaplain until his father's death brought him back to Northampton. There he tried his hand at preaching, but was discouraged to find the local churches virtually empty. Apparently one consequence of revolutionary zeal was neglect of Scripture. After the war ended, Dwight foresaw the dawn of a new era which would restore religious propriety as well as "subordination and obedience to law" and put a stop to the country's "spirit of licentiousness." To impart moral guidance he returned to the pulpit, in affluent Fairfield, Connecticut.[1] But it was not until the forty-three-year-old Dwight was offered the presidency of Yale in 1795 that he finally found the proper forum for his teachings. Infected by the freedom-loving fervor of the day, the college had reverted to its earlier cavalier ways. His students — most of them aspiring clergymen[2] — had devolved into unruly free spirits,

---

Yale University Press, 1989), 40-1.

[1] The post in Fairfield was one of the best-paid in New England at that time. Imholt, "Timothy Dwight," 386.

[2] In 1801, Dwight reported that over the preceding 70 years Yale had graduated

with some adopting the names of prominent French and English "infidels" in a gesture of revolutionary fraternity. The new president immediately set out to restore discipline and religion to their proper places. Instead of forcing students to obey rules, he invited them to consult their consciences. He debated them on the question of "infidelity" and won them over by power of his arguments. Soon Yale was not only attracting many more students, it was also undergoing a religious renaissance.

Though primarily a man of faith, Dwight was passionately interested in the natural world and founded the Connecticut Academy of Arts and Sciences, for the purpose of gathering information on the state's flora and fauna. Like naturalists of his day, Dwight sought proof of divine purpose in the creation of so many varied forms of life: cataloging and classifying species was a way of paying homage to the glory of God.[1] Those who questioned divine wisdom or sought to evade its consequences were committing sacrilege. Thus, Dwight opposed vaccinations against smallpox, arguing that "If God had decreed from all eternity that a certain person should die of [this epidemic], it would be a frightful sin to avoid and annul that decree by the trick of vaccination."[2]

Like John Winthrop and his own ancestors, Dwight also considered any tampering with the "natural order" in political life as going against divine will. Men of high moral character, sound judgment, prudence, and piety should be accepted as superiors in all fields — as governors, poets, military officers, doctors, lawyers, judges, merchants, and land owners. Hence, he had no compunction about taking many different posts — leader of the Federalist Party in Connecticut, head of the state's Congregational Church, and president of Yale — earning Dwight the sobriquet of "Pope of New England." Some of his critics accused him of not only wanting to impose the strict Puritan morality of his maternal grandfather, Jonathan Edwards, on the college, but also of seeking to control American higher education. They mocked his pomposity and penchant for donning "papal" attire in the pulpit.[3] But they also feared Dwight would use his power to bring Connecticut under the college's control.[4] These charges indicate how

2,562 students; of these, 786 had been ordained as Congregationalist ministers. Timothy Dwight, *A Discourse on Some Events of the Last Century, Delivered in the Brick Church, New Haven, on Wednesday, January 7, 1801,* (New Haven: Ezra Reed, 1801), 15.

[1] This impulse motivated the eighteenth-century Swedish botanist Carl Linnaeus, for instance.

[2] This statement is widely attributed to Dwight — most recently in Christopher Hitchens's *God Is Not Great: How Religion Poisons Everything,* but the author was not able to find any original source for it. Dwight, of course, had lost his sight as a consequence of having been inoculated.

[3] Imholt, "Making of Timothy Dwight," 389, 392.

[4] The governor and other high-ranking state officials sat on the Yale

successfully Dwight had fused ecclesiastical, academic, and political power within his own person.

He played a pivotal role in sustaining Federalist-Congregationalist domination in the Nutmeg State until the end of the War of 1812. More than any governor or clergyman, Dwight held the line against more religious or political liberalism. He condemned Universalism and its adherents and strove for a Calvinist religious revival. He published articles and gave speeches advocating Federalist policies. He abhorred "infidels" in public life and prophesied that the election of Thomas Jefferson would "ruin the Republic." [1] He restored academic rigor, personal discipline, and religious observance at Yale and made it the biggest college in the United States.[2] But, formidable as it was, Dwight's career could not withstand the pressure for change in Connecticut, intensified by the existential crisis of the "year without a summer."

New political thinking had already gained ground in 1800 when a wave of populism brought Thomas Jefferson into the White House and dramatically altered the balance of power in the House and Senate.[3] Many states, particularly new ones west of the Appalachian Mountains, began to implement policies which expanded individual rights, including the suffrage. Newly-admitted Vermont became the first state to do so, granting all men over the age of twenty one the vote. New Jersey, Maryland, and South Carolina soon followed suit. Congress also removed the property-owning requirement for the Old Northwest in 1808. This broadening of the suffrage reflected a growing belief that wealth and status should not determine who made the laws. Greater prosperity and the dissemination of information through newspapers supported this national trend.

In New England, Vermont, New Hampshire, and Rhode Island voters favored opening up the political process, electing several Democratic-

---

Corporation's board of trustees, which some likened to the College of Cardinals.

[1] Imholt, "Dwight," 403-07. In an 1801 address, Dwight thus described "infidels" in France: "Emboldened beyond *every* fear by this astonishing event [the French Revolution], Infidelity, which anciently had hid behind a mask, walked forth in open day, and displayed her genuine features to the sun. Without a blush she now denied the existence of moral obligation, annihilated the distinction between virtue and vice, challenged and authorized the indulgence of every lust, trode [sic] down the barriers of truth, perjured herself daily in the sight of the universe, lifted up her front in the face of heaven, denied the being, and dared the thunder, of the Almighty." Dwight, *Discourse*, 28.

[2] Brooks Mather Kelley, *Yale: A History* (New Haven: Yale University Press, 1999), 142.

[3] The Senate went from being two-thirds Federalist in 1798 to forty one percent in 1800. In the House, the shift was from fifty-three to thirty-seven percent Federalist.

Republicans to the governor's mansion and Congress during the first decade of the nineteenth century.[1] But Bay Staters did not put a Democratic-Republican candidate in the Senate until 1810. And only Federalists represented Connecticut in the upper chamber for another eight years. Democratic-Republicans held the governor's office in Massachusetts for just six years during the first quarter of the nineteenth century.[2] Connecticut voters first elected a gubernatorial candidate from this party in 1817. Nutmeg State voters did not elect a single Democratic-Republican to the House before 1816.

Several factors account for these two states being so slow to shift their party allegiance. Tradition was clearly one. So was the Puritan-based respect for one's "betters." Another reason was the clout of merchants and businessmen, who favored protective taxes and tariffs and opposed Western expansion as detrimental to New England's well-being. Given these realities, why did support for still entrenched Federalists disappear so rapidly between 1816 and 1818? In 1816, New England had three Federalist governors and three Democratic-Republicans; a year later, all six were Democratic-Republicans. In 1816, New England had eight Federalist Senators, but, by 1818, only six. The House representation from New England diminished from thirty two Federalist seats to twenty during this two-year period.

To understand this sudden downfall of the Federalists, one needs to examine closely the political circumstances of the day — the predisposing factors. Most historians have concluded that the party brought about its own demise by opposing the War of 1812 and then holding secret talks afterwards about seceding from the Union. These actions cast the party in a decidedly unpatriotic light. But the conflict with England was widely unpopular in the Northeast, because it interfered with lucrative transatlantic trade. When the Federalist governors of Connecticut and Massachusetts refused to send their militias to help defeat British forces, they were not going against public opinion in their states. So it does not seem logical that voters turned against the party for having held this antiwar position. And, while some Federalist politicians did attend an 1814 conference in Hartford, where some argued for leaving the Union because of foreign policy differences with Washington, it seems unlikely that the party itself would have been punished at the polls for this subversive talk.

A more plausible explanation is that New England voters, shaken by the crop failures of 1816, abandoned the party in power to express their anger

---

[1] Vermont and Rhode Island voters first elected a Democratic-Republican to the Senate in 1801; New Hampshire voters first did so in 1805. Vermont had its first Democratic-Republican governor in 1807, New Hampshire in 1805, and Rhode Island in 1807.

[2] During some periods, the governor did not belong to any political party.

and dismay over this blow to their way of life. (It should be noted that there is no written evidence supporting this contention, but human motivation is rarely well documented.) While an irrational response, it would not have been unprecedented: often in the past, threats to food security have led to political upheaval. In fact, this reaction occurred after the disastrously small harvests in Europe that year: rioting and other protests broke out against governments which had not done enough to alleviate the ensuing hunger. Some historians argue that voters in the United States used the ballot to express the same frustration.[1] Nationally, this could explain the wholesale purge of incumbents in Congress in the fall elections: more than two-thirds of House members lost their seats.[2] The big winner was the Democratic-Republican Party, which picked up twenty seven seats, while the Federalists lost twenty five.[3] The latter were closely identified with the standing order and the belief that one could depend upon time-honored ways to provide security as they had in the past. But in 1816 this trust had been shaken. God had not taken care of believers. Because religious faith and political loyalty were so closely entwined in much of New England, doubts about one pillar of certitude were apt to impugn the other: the notion that a wise few men should rule over the many.

***

The most divisive issue faced by the fledgling American republic prior to the "year without a summer" was going to war against England.[4] The impressments of some nine thousand American sailors by British warships, coupled with England's enforcing a ban on all transatlantic commerce with its rival and enemy, had angered many Americans who supported free trade. These affronts to national sovereignty and economic well-being seemed

[1] See, for example, C. Edward Skeen , *1816: America Rising* (Lexington: University Press of Kentucky, 2003), 16. Skeen opines: "The outrage of the citizenry over this act was undoubtedly fanned by the general malaise created by crop failures and threatened famine." The fact that individual states voted for their representatives at different points during the year — for example, New York in April, Connecticut in September, and New Jersey in November — makes its hard to argue that the destructive weather played a role everywhere in the country.

[2] However, most historians attribute this defeat to legislators having voted to double their salaries earlier that year.

[3] But it was not only Federalists who faced the voters' wrath. Democratic-Republican Henry Clay, Speaker of House, narrowly carried his district because of his vote for the Compensation Act. See Robert V. Remini, *Henry Clay: Statesman for the Union* (New York: Norton, 1991), 147. But Clay's Federalist opponent, John Pope, was also unpopular for having opposed the War of 1812. See Clay, *The Papers of Henry Clay*, vol. 1, *The Rising Statesman, 1797-1814*, ed. James F. Hopkins (Lexington: University of Kentucky Press, 1959), 254.

[4] The vote to go to war in 1812 was the closest in American history. At the outset, the United States had only seven thousand men under arms. George L. Clark, *History of Connecticut* (New York: G. Putnam's Sons, 1914), 330.

to demand a military response. But others wanted to avoid a conflict with the former colonial power. Some conservative Federalists still had a strong affinity for their mother country, perhaps even stronger than their ties to their adopted one. Back in 1793, Timothy Dwight had vowed: "A war with Great Britain we, at least in New England, will not enter into. Sooner would ninety-nine out of a hundred separate from the Union, then plunge ourselves in such an abyss of misery."[1] This breach over the war fell along regional and party lines. Those with a vested interest in trade with England — merchants, shipbuilders, shopkeepers, and the like — did not want to take up arms against their major commercial partner. Most opposition was concentrated in New England, which stood to lose the most from protracted hostilities. The Southern and Western states were more eager to assert American sovereignty — and seize some British territory in the process. Thus, when Congress voted on the war resolution, only five Senators and seventeen Congressmen from Northern states — versus sixty two of those from south of Delaware — voted "Aye." The two parties also split over this issue: not one of the thirty nine Federalists in Congress favored the war resolution. So upset were New England voters over the Democratic-Republicans for taking the country to war that they threw many of them out of Congress. [2] Their unhappiness only increased when the U.S. government levied new taxes to pay for raising and equipping an army to prosecute "Mr. Madison's War." [3] In Congress, Federalist leaders denounced the conflict as politically motivated. They objected to attacking Canada, which they considered a friendly neighbor as well as a close trade partner. By warring with England, the United States was also indirectly aiding Napoleonic France — a nation whose revolutionary lawlessness and moral "depravity" appalled many Federalists, especially the clergy. Some leading Congregationalists went so far as to interpret the war as God's rebuke to the Republic for having abandoned its piety.

For these reasons, the Federalist governors of Massachusetts, Connecticut, and Rhode Island rebuffed President Madison's request to call up their militias.[4] They justified their non-compliance on legal grounds: only they, not the President, had the power to mobilize these forces.[5] The

[1] Quoted in William Plumer, Jr., *The Life of William Plumer*, ed. A. Peabody (Boston: Phillips, Sampson, 1857), 283.
[2] In the House, twenty New England congressmen opposed the war declaration, and twelve approved it.
[3] Henry Cabot Lodge, "A Compendious History," vol. 1, *The New England States: Their Constitutional, Judicial, Educational, Commercial, Professional and Industrial History*, ed. William T. Davis (Boston: D.H. Hurd, 1897), 25.
[4] Ronald R. Hickey, *The War of 1812: A Forgotten Conflict* (Urbana: University of Illinois Press, 1989), 260.
[5] Josiah G. Holland, *History of Western Massachusetts: The Counties of Hampden, Hampshire, Franklin, and Berkshire* (Springfield MA: S. Bowles, 1851), 327.

legislature in Boston adopted a resolution complaining that the war was being "waged without justifiable causes and prosecuted in a manner which indicates that conquest and ambition are its real motives . . ."[1] Rhode Island's General Assembly appointed a committee to decide if the state's adoption of the Constitution should be nullified. Clergymen like Lyman Beecher, a protégé of Timothy Dwight's since his days at Yale, feared a federal "military despotism" and felt that independent militias were the only safeguard against its being created.[2]

Federalist reluctance to take up arms evaporated in the middle of 1813, when British warships appeared menacingly off the New England coast. In June, the new governor of Connecticut, John Cotton Smith, dispatched his state's militia to protect American vessels anchored at New London. Soon the British had blockaded most of Long Island Sound. War was suddenly off New England's shores. A few months later, British troops began harassing coastal settlements in Maine. The following spring, enemy warships sailed close to several ports along the Massachusetts coast, choking off ocean-going commerce and crippling the region's economy.[3] Unable to export their produce, thousands of farmers, particularly in Vermont, fell deeply into debt.[4] Countless other New Englanders lost their jobs, and their families were impoverished. Fearing more incursions in the southern part of Massachusetts, Governor Strong finally ordered the militia put on stand-by later that year.[5] But, by staying on the sidelines so long, New England had alienated other sections of the country, coming across as indifferent to the common good. From the New Englanders' perspective, the federal government had abandoned them by invading Canada instead of helping to ward off British troops in their backyards. The lesson they had learned was that it was up to them to take care of their own defense.

The War of 1812 *was* a seminal test of American unity. With the fighting not going well for American forces, it seemed that the test would end badly: the fragile young nation might break apart under pressure from these political crosscurrents. In some corners, hostility to Washington bordered

---

[1] Quoted in Otis, *Letters*, 84.
[2] Lyman Beecher, *Autobiography, Correspondence of Lyman Beecher, D.D*, ed. Charles Beecher, vol. 1 (New York: Harper and Brothers, 1864), 266.
[3] New England's fishing and whaling industries were also devastated. Unable to go to sea, ships let their crews go, causing severe economic distress in places like Nantucket. Russell, *Long, Deep Furrow*, 146.
[4] Samuel Swift, *History of the Town of Middlebury, In the County of Addison, Vermont* (Rutland VT: Charles F. Tuttle, 1971), 94. In Addison County, a wheat-growing area in the western part of the state, more than five hundred farmers were taken to court for non-payment of debt two years after the war ended.
[5] Governor Jones of Rhode Island had reversed his opposition to calling up the militia earlier in the conflict.

on sedition. More and more citizens had come around to the Federalist view that the war was ill-conceived and unjustified. Consequently, as had previously happened in 1808, this party received a boost at the polls in 1814, with solid majorities in several state legislatures.[1] Thus reenergized, delegates from the five New England states — almost all of them anti-war Federalists — convened in Hartford in the middle of December. Along with denouncing a "ruinous" war, the assembled politicians skewered the Madison administration for overstepping its bounds — ironically so, given the Federalists' usual preference for a strong central government. Hotheads who had talked about seceding from the Union had been shooed away from this gathering, so the measures ultimately adopted were less radical. They included Constitutional amendments to do away with the infamous three-fifth rule (which gave slave states extra representation in Congress), prohibit embargoes of over sixty days' duration, and require two-thirds majorities in Congress to go to war and to admit new states to join the Union, thus preserving New England's power at the federal level.[2]

But, of course, it was *Federalist* power they were bent upon maintaining. America's ill fortune in its war with England had set the stage for the party of John Adams to claw its way back to power, by laying blame at the feet of the Democratic-Republicans. But there were several flaws in the Federalist strategy. First, not all New Englanders had opposed the war: the region had, in fact, supplied more troops than any other region to fight it.[3] Secondly, while the delegates in Hartford were deliberating, some twelve hundred miles away, south of New Orleans, American forces commanded by Major General Andrew Jackson soundly routed the British on January 8, 1815, turning a looming defeat into what was misconstrued as a tide-turning victory. Then, two weeks earlier, on 24 December 1814, at Ghent, Belgium, representatives of the English and American governments had signed a peace treaty restoring the status quo ante between their nations. News of these events had not yet reached Hartford when the convention disbanded on 5 January 1815. So when

---

[1] In Rhode Island and New Hampshire, Federalists held on to their narrow majorities in the legislature. Massachusetts voters gave this party an overwhelming victory in 1814, with a margin of 203 seats in the assembly and a nearly two-to-one majority in the state senate. Connecticut Federalists similarly maintained legislative control comfortably. While losing some twenty seats in the April 1815 legislative election, the Federalists still outnumbered the Democrats, 143–57. See *Public Records of the State of Connecticut, vol. 17, May 1814 through October 1815*, ed. Donald M. Arnold (Hartford: Office of the State Historian, 2000), 295. Only Vermont, a Democratic-Republican bastion, went against this electoral tide in New England.

[2] David S. Muzzey, *An American History* (Boston, New York: Ginn and Co., 1911), 224.

[3] New England contributed a force of nineteen regiments, compared to only fifteen from the Mid-Atlantic states, and ten from the Southern ones.

a group of delegates reached Washington a few days later, bearing anti-war resolutions in their saddle bags, they were dumbfounded to find the capital riotously celebrating America's "triumph."[1]

Thirdly, the Federalists lost support by helping to pass the highly unpopular Compensation Act of 1816. Designed to allow their pay to keep pace with inflation, this bill awarded members of Congress an annual salary of fifteen-hundred dollars, replacing their previous, hourly remuneration.[2] As anticipated, such feathering of their own nest, at a time when most Americans were struggling, created a firestorm of outrage and protest — a new phenomenon in the United States.[3] House members were burned in effigy, as the public turned on elected officials en masse. During the Congressional elections later that year (and the next), nearly seventy percent of incumbents were defeated — the largest rebuke ever experienced by members of Congress in a single electoral cycle. [4] In New England, only four of the twenty seven office holders who had voted for this act were re-elected; in the Mid-Atlantic states, six out of twenty two.[5] Because a higher percentage of their members of Congress had backed this bill, the Federalists took the brunt of this outrage: only nine of the fifty two members of their caucus returned to Washington in 1817.[6] Realizing the Compensation Act was a colossal blunder, Congress quickly repealed it. But some of the damage was irreparable. Many historians have argued that this self-serving legislation was the final nail in the coffin of the Federalist Party: no longer would voters tolerate a party so selfishly preoccupied with its own financial concerns.

---

[1] When he first heard of the English attack on New Orleans, in mid-January, The Reverend Robbins had written in his diary: "I hope the British will take it." Robbins, *Diary*, entry for 16 January 1815, 616.

[2] Since 1789, a congressman had been paid six dollars a day. William T. Bianco, David B. Spence, and John D. Wilkerson, "The Electoral Connection in the Early Congress: The Case of the Compensation Act of 1816," *American Journal of Political Science* 40:1 (February 1996): 145.

[3] This was fueled by the press. According to one study of this legislation, editors were eager for a big news story to attract readers and "served up" the Compensation Act for months as a "standing dish . . . boiled, roasted, and fricasseed." Bianco et al., "Electoral Connection," 145.

[4] House campaigns stretched from April 1816 to August 1817. The exact figure was 68.7 percent. Bianco, et al., "Electoral Connection," 145.

[5] Skeen, "'Vox Populi, Vox Dei': The Compensation Act of 1816 and the Rise of Popular Politics," *Journal of the Early Republic* 6 (Fall 1986): 266.

[6] Nationally, only fifteen of the eighty-one members of Congress who voted for this bill were re-elected. Ibid., 266. Furthermore, Federalists had supported the Compensation Act more than Democratic-Republicans had: 64.4% of Federalist congressmen voted in favor, versus only 38.2% of the latter party. Bianco, et al., "Electoral Connection," 145. This article contends that Federalists were more ideologically inclined to vote for a pay increase. Many Federalists who voted for the bill retired after its passage, fearing voter retaliation.

But there are several problems with this interpretation. If the war was such a wedge issue, why did the Federalists not incur the wrath of voters *earlier* for their unwavering opposition to it? Clearly this was because so many Americans — chiefly New Englanders — shared the Federalists' position. They, too, thought the war was a mistake. Or, even if they disagreed with the party on this issue, they were not inclined to express this dissent at the polls. In the Congressional elections held in 1814 and 1815 — more than two years into the war — the Federalists *gained* two seats in the Senate and lost only four in the House. This outcome suggests that the conflict with England was not a significant "tilt" factor, one way or the other. While the euphoria inspired by Jackson's victory at New Orleans may have swayed many to rethink their position, the underlying reasons for opposing the war — protecting New England's economic interests and political dominance — had *not* changed. The same analysis would seem to rule out the Hartford Convention as the precipitating factor for the demise of the Federalists. True, its rumblings about secession stuck many as treasonous. [1] But assertions of states' rights were then — as now — greeted with approval in many quarters only three decades after the Thirteen Colonies had voluntarily given up their sovereignty to form a nation. There is no way to measure support for the principles of the Hartford Convention — there were no opinion polls in those days — but anecdotally it appears to have been strong. The Massachusetts state legislature was in favor, and its views generally reflected those of the voting public.[2] This body was controlled by the Essex Junto — a group of pro-British Federalists who had banded together shortly after the Revolution to prevent the "over-rapid advance" of democracy in the state.[3] In other words, the traditional elites still ran public affairs in Massachusetts, and voices favoring political change were still largely silent.

But the outcry over the Compensation Act suggested that a new force was emerging: public opinion. Voters had leveled their wrath at elected officials for dipping into the public coffers so that they could live more comfortably. This was a landmark election in American history, indicating, as the historian Edward Skeen has pointed out, the end of the era of "deferential politics," in which voters passively accepted their elected officials as their betters and went along with what they proposed. The notion that the will of the

---

[1] Andrew Jackson was prominent among those who labeled the delegates "traitors." Clark, *History of Connecticut*, 337.
[2] Otis, *Letters*, p. 10-11. A motion in favor of the convention passed with large majorities in the state legislature. The vote in the Massachusetts House was 226 -67 in favor, and in the Senate, 22-10.
[3] Morse, *Federalist Party*, 16, 18. The junto had convinced a majority of Massachusetts voters to reject a state constitution proposed in 1778 by the legislature because it went too far in enshrining more individual rights in Massachusetts.

electorate was supreme and had to be heeded was superseding blind faith in higher authority.[1] In New England in particular, an English aristocratic system of governance was evolving into a form of democracy. In the span of just a few years, the promise enshrined in the Declaration of Independence began to be realized.

Several developments helped to bring about this radical change. As mentioned earlier, migration to the Western frontier was one. The War of 1812 was another milestone in this transformative process. By standing up to the armies of a powerful England, Americans gained confidence in themselves and their form of government. They earned the right to be treated as an equal among equals. Because England was a monarchy, America's "victory" also demonstrated that the ordinary people of a democracy could best a European "tyrant" and his minions.[2] Much as the British defeat of the French emperor at Waterloo had stimulated longings for freedom in Europe, so did a similar outcome in North America stoke an appetite for democracy and self-reliance.[3] Vilification of the defeated French ruler became the mantra of the day. By implication, all leaders who opposed freedom were the enemies of the people — and of God. For example, in a speech celebrating the triumph over Napoleon, on 9 September 1814, Maryland lawmaker John Hanson Thomas proclaimed, "The arm of heaven has been, at last, visibly stretched forth in pity to release Christendom from servile terror and impending bondage. The thunder of retributive justice has rolled over the oppressor. The lightning of its vengeance has struck the iron crown from his guilty head. . . . The ways of God to man are vindicated on earth. Man is again free. The earth rejoiceth and all the ends thereof are glad."[4]

Using similarly apocalyptic rhetoric, Elias Smith, a populist preacher and editor of the *Herald of Gospel Liberty*, linked religiosity and freedom in hailing the end of Napoleon's suzerainty. Genuine righteousness, he affirmed, can

---

[1] Skeen, *America Rising*, 77.

[2] According to historian Alan Taylor, "Most voters preferred the comforting myth of a glorious war confirmed by an honorable peace," and thus abandoned the Federalists in drones. In the congressional elections of 1816, the party lost a third of its seats. Taylor, *Civil War of 1812*, 421.

[3] For a discussion of the impact of Napoleon's defeat on the political climate in Europe, see Benedetto Croce, *History of Europe in the Nineteenth Century*, trans. Henry Furst (New York: Harcourt, Brace and World, 1963), 3-16. Many prominent figures in American public life and letters were inclined to see parallels in the outcomes of these two wars. See, for example, Henry Ware, Jr., *A Poem Pronounced at Cambridge, February 23, 1815, at the Celebration of Peace between the United States and Great Britain*, (Cambridge MA: Hilliard and Metcalf, 1815).

[4] John Hanson Thomas, "Oration Delivered at the United Celebration at Shephardtown, on the Potomac, July 28th, 1814, on the late glorious events in Europe," *Burlington Gazette*, 19 September 1814. The Reverend Thomas Robbins described Napoleon's defeat at Waterloo as caused by the "merciful interposition of heaven." Robbins, *Diary*, entry for 5 August 1815, 636.

only exist when kings and popes no longer wield power. "Monarchies, and hierarchies will be driven from the hearth, for God has declared it by his prophets. A Republican government has proved . . . its strength to withstand all the power of the British lion, or the pretended mistress of the ocean." The task now facing a "righteous" nation was to complete the work of the Revolution, to assure that the "yoke of tyrants" never again fell upon the shoulders of the American people.[1] Even before Americans took up arms against the British, a hankering for more freedom had been evident throughout the land, and not only in Republican circles. In 1811, for example, Connecticut lawyers affiliated with the Federalist Party rejected the clergy's nominee for governor, incumbent John Treadwell, in favor of Roger Griswold, considered the "ablest man in Congress."[2] And, in Massachusetts in 1814, a band of young Federalists had likewise refused to accept the House candidate of the party's elders and put forward their own choice.[3] A third rebellion against the establishment came in December 1814, when a group of Maine leaders, angry at Massachusetts' failure to defend their coastline, proposed that the District should break away and become an independent state.[4]

Over the course of several decades, these various events advanced the democratization of American society. As happens frequently in human history, a long-developing trend can suddenly gain momentum and intensify — "swerve" in a new direction, much as occurred in the course of Western thought when Lucretius' major poem, *De rerum natura*, was rediscovered and widely copied by a worldly, book-loving monk in the early 15th century.[5] The failed harvests and hardship caused by the extremely cold spring and summer weather in 1816 arguably had this effect. Someone had to be blamed for this calamity, and higher officials are the easiest to scapegoat — even if they bore no real responsibility over what had happened.[6] Previously, tens of thousands of peasants and city dwellers alike had vented their anger over

---

[1] Elias Smith, "Peace! Peace!" *Herald of Gospel Liberty* (Portsmouth NH), 3 March 1815.

[2] As a result of this rare break in Federalist ranks, Griswold was elected that year. Beecher, *Autobiography*, 259.

[3] A Federalist caucus in Boston was forced to come up with a third, compromise candidate to resolve this dispute. Samuel Eliot Morison, *The Life and Letters of Harrison Gray Otis, Federalist, 1765-1848*, vol. 1 (Boston: Houghton Mifflin, 1913), 295.

[4] Banks, "War of 1812," 164.

[5] For this fascinating story, see Stephen Greenblatt, *The Swerve: How the World Became Modern* (New York: Norton, 2011).

[6] The argument that the drought and cold weather of 1816 had political consequences has been advanced most notably by Edward Skeen. See, for example, Skeen, "Vox Populi, Vox Dei," 267-8. Of course, in the absence of public-opinion polls, it is impossible to determine what factors influenced voters' decisions in the 1816-1817 congressional elections.

bread shortages caused by drought and cold weather in France in 1788–1789, and these demonstrations had helped to foment a revolution. A quarter century later, as was noted above, the prospect of famine sparked more protest, social unrest, and violence in several European countries.

As was the case during such other major crises, the "year without a summer" underscored that providing for the security and well-being of a nation required leaders to take more steps other than beseeching God for aid and comfort. Drought, lack of food, and starvation were not remedied by such imprecations. In the early nineteenth century, continued reliance on the Almighty, as practiced by the Puritans of old, seemed inadequate and ineffectual. The growth of government, with its legitimacy deriving from the consent of the government, raised expectations that problems of this nature could — and should — be solved. The view that forces controlling the natural world were divinely governed and thus humanly incomprehensible was giving way to a curiosity about how such laws operated. The influence of the Enlightenment was making itself felt. Where once Pilgrim families, huddled around a spitting hearth, had passively accepted the roar of winter storms howling at their windows as a test God expected them to pass to prove their faith or atone for their sins, educated men now contemplated alternative explanations.[1] Some, like the voraciously inquisitive Thomas Jefferson, noted a new pattern in the weather — the winters were growing milder — and speculated that God might not be responsible. Figures as disparate as Daniel Webster (who drew on meteorological records) and Cotton Mather (who believed the Deity made all happen according to His plan) concluded that this warming trend was likely caused by the clearing of New England's forests.[2]

This investigatory approach to the physical world diminished the power of the clergy, who no longer had to be consulted in the search for truth. New England's Congregational ministers were already losing their godlike monopoly in this quest because of a growing diversity in religious life. In Connecticut, Quakers and other dissenting sects gradually began to exercise their right to hold worship services. Slowly Congregational parishioners slipped away to other houses of worship. In 1760, Connecticut had only nine Baptist churches, serving a mere 450 members, but half a century later this denomination had established sixty-five churches and encompassed some five thousand seven hundred persons. Eight years later, the Baptists

---

[1] Bernard Mergen, *Snow in America* (Washington DC: Smithsonian Institution Press, 1997), 4.

[2] Ibid., 1. See also Cotton Mather, *The Christian Philosopher: Collection of the Best Discoveries in Nature with Religious Improvements* (Charlestown MA: Middlesex, 1815). Originally published in 1721, this latter work reveals the extent of Mather's interest in, and knowledge of, recent scientific discoveries, which only deepened his belief in divine causality.

could boast eighty-five churches and eight thousand members. By then, the dissenting churches — taken together — were greater in number than the Congregationalists'.[1] The era of Calvinist theocracy was over. After 1740, the Great Awakening had spread across New England, emphasizing individual conversion over submission to clerical control. "Rough-hewn," self-trained, and self-appointed ministers were now preaching the word of God, without degrees from Yale or years of biblical exegesis.[2] As a traditionalist like Lyman Beecher saw it, this evangelical movement aroused prejudice toward the Congregationalists and encouraged a splintering of the Christian community, as well as an unwelcome enthusiasm "which defied restraint and despised order."[3] Revivalists freely debated what the role of a church should be — acknowledging that ordinary folk had as much to contribute to this debate as anyone else.[4] They were also free to interpret the Bible for themselves.[5]

The rise of "infidel philosophy" — a derogatory term applied loosely to Deism and Universalism, Descartes and William Ellery Channing — further undermined the Calvinists' position. Making it easier to become a member of the Congregational church (in order to boost membership) through a practice known as the "half-way covenant" enabled persons to attend services and have their children baptized without requiring them to lead morally upright lives.[6] This policy took away a minister's power as arbiter of piety. More generally, some clergy came under critical scrutiny for their extravagant lifestyle, imperious manner, and British affectations. The unstructured and egalitarian worship services introduced by the Great Awakening made these clerical mores appear antiquated — and out of step with their parishioners' values. Unitarian leaders like William Ellery Channing linked the Congregationalist Church with a much-despised "papal tyranny" in Rome. In an 1819 sermon he lambasted the Calvinists for preaching that only a chosen few would be saved, for perpetuating a "gloomy, forbidding, and servile" religion, and for concealing "under the name of pious zeal the love of domination, the conceit of infallibility, and the spirit of intolerance, and trampling on men's rights under the pretence [sic] of saving their souls."[7]

---

[1] William G. McLoughlin, *New England Dissent, 1630-1833*, vol. 2, *The Baptists and the Separation of Church and State* (Cambridge: Harvard University Press, 1971), 919.

[2] Nathan O. Hatch, *The Democratization of American Christianity* (New Haven: Yale University Press, 1989), 14.

[3] Beecher, Lyman Beecher, Sermon IV: "The Building of Waste Places," Wolcott CT, delivered 21 September 1814, *Beecher's Works*, vol. 2, *Sermons Delivered on Various Occasions* (Boston, Chicago: John Jewett, 1852), 119.

[4] Hatch, *Democratization of Christianity*, 9-16.

[5] Ibid., 141.

[6] Beecher, *Beecher's Works*, 117-18.

[7] William Ellery Channing, "Discourse at the Ordination of the Rev. Jared Sparks, Baltimore, 1819," *The Works of William E. Channing, D.D.*, vol. 3, 11th ed.

At the end of the War of 1812, this clamor for greater democracy was growing louder. Having embraced institutions built on merit rather than bloodlines, America was poised to enter a new era of growth and prosperity. Only New England was holding back. There an intractable Federalist Party clung to power, winning several statewide elections in 1814 despite having backed Congressional pay increases and entertained thoughts of leaving the Union. Its departure from American political life was neither imminent nor inevitable. That would require a further, dramatic loss of popularity. The fact that this occurred immediately after the disastrous "year without a summer" strongly suggests a causal relationship. Disillusioned by blighted crops and the cavalier optimism of the ruling elite, New Englanders were looking for an efficacious response to this crisis. Their unhappiness was evident in the outcome of Congressional elections held around that time: in three New England states, the Federalists lost a total of twenty seats out of the thirty-two they had previously held.[1] The event which best attested to this shift in political alignment was the contest for governor in Connecticut in 1817 — which resulted in the first Democratic-Republican victory in the state's history. This outcome signaled that this redoubt of "steady habits" was adopting new, more tolerant attitudes and values. The era of Congregational-Federalist rule was over. In the Nutmeg State, the American Revolution had finally arrived.

***

On the face of it, the person who won that Connecticut election was an improbable figure to turn the state's politics upside down. His name was Oliver Wolcott, Jr., and he was as steeped in the Federalist tradition as any candidate could be. The oldest son of a former governor (who had signed the Declaration of Independence), and the grandson of another, he was born in Litchfield in 1760 and attended Yale, following in his father's footsteps. During the Revolution, Wolcott served two stints in the army, ending up as a quartermaster in Litchfield. After studying law, Wolcott entered public service, holding several finance-related positions in state government. In 1785 he married Elizabeth Stoughton, who would eventually bear him seven children. During the administration of George Washington, the twenty-nine-year-old Wolcott moved to Washington with his wife and growing family, to serve as auditor for the U.S. Treasury. On the recommendation of Alexander Hamilton, he was appointed U.S. Comptroller in 1791 and then, after Hamilton resigned, became the nation's second Secretary of

---

(Boston: George G. Channing, 1849), 87, 98.
[1] In Vermont and New Hampshire, the make-up of the congressional delegation went from all Federalist to all Democratic-Republican. In Massachusetts, the Federalists lost eight of the sixteen seats they had held before 1817.

---

the Treasury. In November 1800, the month Jefferson was elected to the presidency, Wolcott abruptly resigned this post, irate over Democratic-Republican charges that he had tried to destroy documents purportedly showing that he had embezzled federal funds. This was an early sign of his independent temperament.

With the Democratic-Republicans running the country, Wolcott returned to the private sector, becoming president of the Merchants Bank, in New York. Shortly thereafter he joined with his younger brother Frederick and a Federalist congressman named Benjamin Tallmadge to launch a trade venture with China, importing tea, silk, pearls, and other goods. This import business — which made him a very rich man — lasted for several years. Subsequently, Wolcott returned to banking, as the first president of the Bank of North America. He resigned in 1814 over a dispute with the bank's directors — another instance of his tendency to stand fast by his principles.

Wolcott did not agree with the Federalists' opposition to the war with England. His years in Washington had taught him the importance of national unity in foreign affairs. Thus he could not countenance the parochial and self-serving refusal of New England Federalists to furnish troops, or their idle talk of seceding from the Union. In New York, where he was then living, the war remained a controversial issue in the 1815 elections, with Federalists criticizing the higher taxes the war had made necessary. In response, Wolcott convened a group of fellow dissident Federalists to establish an American Federalist Party. It ran candidates for the state assembly on a pro-war platform, but only garnered one percent of the vote. After this setback, Wolcott returned to Litchfield, to assume the life of a gentleman farmer and set up wool-producing factories with one of his brothers. But the political bug soon bit him again. This time he announced his intention to seek the governor's office in 1816.

Wolcott's illustrious lineage gave him a distinct advantage. However, he faced three major obstacles. John Cotton Smith, a prominent Federalist and son of a Puritan minister named for Cotton Mather, had served in the state legislature, in Congress, and as lieutenant governor before entering the governor's mansion in 1812. For Wolcott to wrest the party's nomination away from him would be an uphill struggle, given Connecticut's history of honoring seniority.[1] Furthermore, because he had spent so many years outside the state, Wolcott would be seen as an interloper. Finally, Wolcott had the reputation of being an iconoclast. To overcome these liabilities, he adopted an ingenious strategy. Instead of contesting the Federalist nomination (and probably losing), Wolcott agreed to run on the ticket of a new "Toleration" party.

---

[1] Ironically, his grandfather, Roger Wolcott, was one of only two Connecticut governors in the eighteenth century to have been removed from office.

This was a loose coalition of splinter groups, united in opposition to the exclusive power and privileges of the Federalist/Congregationalist establishment. As its name implied, this upstart party called for tolerance of unorthodox viewpoints — in both politics and religion. Its name harkened back to the 1689 Toleration Act, which had given dissenting churches the right to exist in the colony.[1] But the word "toleration" also evoked Enlightenment thinkers like Locke and Hobbes, who had objected to the constrictions of orthodox Christian political theology. To dispel this "Kingdom of Darkness," these thinkers had proposed a "Great Separation" between religion and worldly activities, including politics. A "new science of man" was to replace blind belief in the divine.[2]

For Oliver Wolcott, the forces of "darkness" threatening Connecticut were more mundane. Looking from a business perspective, he saw its economy dominated by powerful bankers, who restricted credit to would-be entrepreneurs, thus stifling new businesses, and perpetuating the exodus to the Western states.[3] But this problem was only symptomatic of a larger defect in the state's governance. A handful of men set policies on matters as diverse as who could vote and what was considered a sin.[4] This tightly controlled system discouraged new thinking and kept Connecticut languishing behind the rest of the country.[5] During his years in New York Wolcott had witnessed the dawning of a new era in manufacturing and trade. Now he wanted to introduce the same forward-looking attitude in the Nutmeg State, replacing its "steady habits" with new political and economic freedoms.

In order to win the gubernatorial election, Wolcott had to create a coalition of outside groups which shared this desire for more rights — religious, social, or political. The common thread running through his campaign was "Toleration." The common cause was access to power. The

---

[1] Oscar Zeichner, *Connecticut's Years of Controversy, 1750-1776* (Williamsburg, VA: University of North Carolina Press, 1949), 13.

[2] Mark Lilla, *The Stillborn God: Religion, Politics, and the Modern West* (New York: Knopf, 2007), 75-79, 92,101, 111.

[3] "Camillus," untitled article, *Columbian Register* (New Haven), 8 March 1817.

[4] The historian Sean Wilentz has written that, aside from the "quasi-feudal polity of Rhode Island," Connecticut was the nation's most "undemocratic" state in 1815. Less than half of its adult white males enjoyed the right to vote. Wilentz, *Rise of American Democracy*, 183.

[5] Lacking a state constitution, Connecticut was still governed by Crown laws, which restricted the suffrage to those adult males who owned property worth forty pounds ($134) or more, or whose real estate holdings were assessed at forty shillings ($7) or more. In the 1824 presidential election, only 14.9% of the state's adult male voters cast ballots. Table 2, "Percentage of Adult White Males Voting in Elections," Stanley L. Engerman and Kenneth L. Sokoloff, "The Evolution of Suffrage Institutions in the New World," (February 2005), 35. http://www.econ.yale.edu/seminars/echist/eh05/sokoloff-050406.pdf.

common vehicle was "Republicanism." And the common cry in speeches and the press was "More democracy!" Because of his family's prominence in Connecticut's aristocracy, Wolcott seemed ideally suited to wrest power away from the Federalists: voters would understand he was no radical. In April 1816 he came to close to winning, losing by fewer than fourteen hundred votes out of nearly twenty two thousand cast — the best showing ever by a non-Federalist gubernatorial candidate. Still, Cotton Smith was elected for the fourth consecutive time, and Connecticut remained a Federalist state. Those running on Wolcott's "Toleration" ticket did capture eighty five seats in the General Assembly — more than in any previous state election. This outcome came as a shock to old-line Federalists like Thomas Robbins, who could only find comfort in the fact that a "kind Providence" had assured a Federalist victory.[1] To neuter their opposition, the Federalist-controlled legislature quickly passed several bills granting more rights to dissenting groups.[2] But these measures failed to pacify unhappy Connecticut voters.[3]

Diversity, equality, and inclusion were what the "Toleration" party offered. If its adherents were elected, Connecticut would no longer reserve positions of power exclusively for members of the Congregational Church. Tax revenues would no longer pay only the salaries of Congregational clergy. Public funds would no longer support a Yale College which allowed only Congregationalists to serve on its governing board and faculty. Favoritism in politics as well as religion would be eliminated, and "tolerance" enshrined in a state constitution — a document Connecticut had never gotten around to writing. No more would only property owners be able to vote. Wolcott and other leaders of the party affirmed the suffrage as "one of the greatest civil privileges we enjoy," a "bulwark of our liberty."[4] This message was widely disseminated by highly partisan newspapers such as the *Hartford Times*,

---

[1] Robbins, *Diary*, entry for 13 April 1816, 665.
[2] These included the Act of October 1816 for the Support of Literature and Religion, which allocated portions of the federals funds owed to Connecticut — some $145,000, for expenses incurred during the War of 1812 — to minority denominations in the state. Maria Louise Greene, *The Development of Religious Liberty in Connecticut* (New York, Boston: Houghton Mifflin, 1905), 172.
[3] Rather than welcome these funds, the dissenting churches denounced the Federalist legislators for trying to bribe them into accepting continuance of Congregationalist hegemony in the state. They also complained about Congregationalist Yale University's receiving some of these federal funds. Many members of other Protestant sects in Connecticut felt this act would only perpetuate the incestuous bond between political and ecclesiastical power in the state. Siegel, "'Steady Habits,'" 468. Cf. Greene, *Development of Religious Liberty*, 468, 471. According to Greene, the Toleration ticket consequently garnered all votes cast by Methodists in the 1817 gubernatorial election.
[4] Untitled article, *Republican Farmer* (Bridgeport CT), 27 March 1816.

whose founding editor, John M. Niles, devoted his first issue in January 1817 to touting the virtues of a free press and the Republican principles of "just and equal civil and religious privileges," while attacking the Federalist "faction" for its "narrow and contracted views" and "spirit of bigotry and intolerance."[1]

In the 1817 campaign for governor, the Democratic-Republicans faced uncertain prospects. They had nearly won the year before, but many Connecticut farmers who might have voted for the "Toleration" candidates had since left the state.[2] Nonetheless, the reform movement was still gaining ground. Members of dissenting churches were speaking out against their second-class status and demanding reform. An Episcopalian wrote to the *Columbian Register* to complain that his religious affiliation made him feel like a "wretch of the rankest malice." On the defensive, Federalists attacked their opponents as traitors and demagogues. Firing back, Democratic-Republicans denounced the religious litmus test for justices of the peace, court clerks, judges, and other public officials. As the April election neared, the "Toleration" camp drove home the point that Connecticut's outdated way of governance — exclusion of nearly half of the state's freemen from voting— was making it a national laughingstock.[3] Insisting on preserving a state religion was demeaning to other faiths.

Increasingly worried about the outcome, the Federalists leveled personal attacks at Wolcott and other "Toleration" leaders and plastered cities and towns with circulars raising fears about what a Republican victory might bring.[4] They tried to convince undecided voters that *they* were the only true defenders of the Constitution — an argument undermined by the party's

---

[1] "To the Public," *Times* (Hartford), 1 January 1817. A letter in the *Times* would later attack the Federalists for turning Connecticut into an "outlaw from the American family." "A Republican and Toleration-man," "To the Real Friends of Republicanism and Toleration," *Times*, 1 April 1817.

[2] The situation for Democratic Republicans early in 1817 was described as not at all "inviting" because of the flow of residents out of the state. "For the *Register*," *Columbian Register* (New Haven), 15 February 1817. Federalists encouraged their departure, hoping to improve their fortunes by thus getting rid of these democratic-minded "undesirables." *Peopling of New Connecticut*, 9.

[3] See, for example, "M. Servetus," "The Age of Improvements," *Times* (Hartford), 11 March 1817. See also "Erskine," "For the *Times*," *Times*, 11 March 1817.

[4] "Mercury," "Take Notice," *Republican Farmer*, 5 March 1817. Wolcott was accused of being surrounded by a group of "artful, designing, intriguing men, whose views have little regard to religion, or religious toleration." "An Old Freeman," "To the Freemen of this State," *Connecticut Courant*, 25 March 1817. (The editor of the *Courant* was Timothy Dwight's younger brother, Theodore.) Other articles attacked him for "apostasy" for having switched parties and recalled that he had been accused of setting fire to government buildings while Secretary of the Treasury. "Z," "For the *Connecticut Journal*," *Connecticut Journal*, 1 April 1817.

recent refusal to take up arms to preserve the Union.[1] In newspaper editorials, the Federalists dismissed the need for reform: there was no "persecution" in Connecticut, as their opponents claimed. They belittled the "Tolerationists" for being "inconsistent and unprincipled."[2] They predicted that a Federalist defeat would be disastrous for the state: "Connecticut revolutionized, will be the most miserable community to be found among civilized nations."[3] But the fundamental argument that the people of Connecticut did not enjoy equal rights was more persuasive. At the same time, hesitation about abandoning their "steady habits" and turning over power to a new party was waning. Wolcott's character, his image as a "true" Federalist, his Congregationalist affiliation, and service during George Washington's administration were reassuring.[4] So were his pledges to halt the flow of skilled workers out of Connecticut and revive the state's economy by increasing competition in the private sector.[5] More and more people in the state were concluding that the old ways of doing business stood in the way of a better future. Connecticut's "degraded and disgraced situation" had to change.[6] The state had to keep pace with the rest of the nation. It seemed clear that this election would produce an historic outcome, determining whether or not "liberal and high-minded republicanism is hereafter to constitute the prominent feature of the government of Connecticut."[7]

How much of an influence did the dreadful weather of 1816 have on voters' decisions? The lack of polling data makes this question unanswerable. But two consequences of the year's unsettling shifts in temperature likely weighed on their minds as they cast ballots on 7 April 1817. First of all, their sense of security had been badly shaken. What the future might bring was worrisome: it was possible they would not be able to continue living in New England. Changing economic conditions were forcing many to contemplate profound alterations in their own lives, including relocation to the West. In this time of upheaval, new political and religious thinking had more appeal. Their once trusted "wise men" — in government and religious life — did not seem up to the challenges of the day.

After the first devastating frosts, ministers had stood before their congregations and asked for prayers and confession of sins to end the drought. Timothy Dwight had preached that "all creatures wait upon God, that he may give their meat in due season; and that he opens his hand, and

---

[1] "A Federal Churchman," *Columbian Register*, 8 March 1817.
[2] "Make Ready!" *Connecticut Journal*, 1 April 1817.
[3] "To the Freemen of this State," *Connecticut Courant*, 25 March 1817.
[4] "Toleration Ticket," *Times*, 1 April 1817.
[5] "Camillus," untitled article, *Columbian Register*, 8 March 1817. See also "Emigration and Tolerance," *Times*, 8 April 1817.
[6] Untitled article, *Times*, 11 March 1817.
[7] "Laurens," "For the *Times*," *Times*, 11 March 1817.

satisfied the wants of every living thing."[1] But the crops had failed. The Reverend Thomas Robbins had promised the faithful in East Windsor that "whatever you ask in prayer, you will receive, if you have faith." But, by and large, the drought had continued.[2] The Reverend William Bentley, in Salem had insisted in late summer that the prospects for crops were "as good as ever."[3] But many in his state had not had enough to eat. Meanwhile, the Federalist press had downplayed the significance of the cold spells and ridiculed those who left for warmer and more abundant lands. The lesson of the "year without a summer" was that no one really knew what the future would bring. The people had to think for themselves.

When the ballots were tallied, it emerged that the "Toleration" party had won a squeaker. By a margin of 334 votes out of nearly twenty five thousand cast, the voters of Connecticut had turned their backs on tradition and thrown a Federalist governor out of office in favor of a moderate Republican. And this was no one-time fluke. Wolcott would be reelected the following year with more than eighty six percent of the vote and go on to serve as governor for another eight, consecutive, one-year terms, until 1826. No Federalist would ever occupy this office again. The party's fall from power was not a sudden and unexpected event: dissatisfaction with it had been growing for years, as is reflected in the Federalists' shrinking margins of victory in previous gubernatorial elections.[4] Many developments account for this — the decline of the state's economy, the national trend toward republican government, the obsolescence of Connecticut's institutions, the Federalists' "unpatriotic" stance on the War of 1812, their secessionist conniving, and the failure of the party's leaders to respond to changing circumstances. The disastrous weather of 1816 put their deficiencies under a harsh light.

Connecticut's 1817 election pitted two starkly contrasting philosophies against each other. The underlying question was, "Do God and His self-anointed spokesmen on earth control human destiny, or does it fall to citizens to act in their own best interests because no higher power can be expected to protect them?" Or, as the *Hartford Times* expressed it, would Republicanism — "a principle implanted by the God of nature in every human bosom," supersede the "frozen and wintry region of intolerance" which had long held humanity in "unrelenting bondage"?[5] The outcome affirmed that the Nutmeg

---

[1] Dwight, Sermon XIII, "Wisdom of God," *Theology Explained*, 228.

[2] Robbins, *Diary*, entry for 4 August 1816, 675.

[3] Quoted in Barrows Mussey, "Yankee Chills, Ohio Fever," *New England Quarterly* 22:4 (December 1949): 442.

[4] Smith had captured 72.8 per cent of the vote in 1814, but only 59.2 per cent the following year and 52.7 percent in 1816.

[5] Untitled article, *Times*, 25 March 1817. Wrote a "Friend to the Minority Sects"

State was entering a modern era of governance, based upon written law and popular will, rather than the noblesse oblige of its hereditary elite.[1]

Connecticut was not alone among the New England states in undergoing this political transformation. A month earlier, New Hampshire voters reelected the Republican governor, William Plumer, and gave his party majority control of the state's two legislative chambers. And in Rhode Island, the Republican candidate Nehemiah Knight narrowly defeated the Federalist incumbent in April, while his party captured the upper house of the state legislature.[2] This left the venerable party of George Washington still in power in only one governor's mansion in all of New England — in Massachusetts.[3]

On a cold January day in 1817, Timothy Dwight, who had long suffered from prostate cancer, finally succumbed to that disease at the age of sixty four. His funeral in New Haven drew a huge throng of the faithful as well as fellow clergy from all over the state. Local shops and businesses closed their doors in his honor.[4] Amidst an outpouring of grief, the defender of the old faith of Jonathan Edwards and the "steady habits" of his Puritan ancestors — the intractable foe of "infidels," natural law, European immigrants, democracy, and Enlightenment thinking — was laid to rest in the Grove Street Cemetery near his beloved Yale. Dwight's hand-picked choice to succeed him, Jeremiah Day, was a professor of mathematics and natural philosophy at Yale. Committed to maintaining the high standards, religious values, and "classical curriculum" his illustrious predecessor had developed at the college, Day would also turn out to be a more secular, nineteenth-century president than Dwight might have imagined. In the ensuing years Day would add classes in chemistry, mineralogy, geology, political economy, and geography to Yale's course of study. He would urge students to think critically instead of merely memorizing the materials they were presented. They were to apply inductive reasoning and consider evidence before reaching any conclusions. Less than

---

in the *Columbian Register*, "The people are shewing [*sic*] a disposition no longer to bend their neck to the Clergy of the standing order; to think in religious and political matters or [*sic*] themselves; to determine and act without fear, without arbitrary force." "Whining," *Columbian Register*, 5 April 1817.
[1] A eulogy for Smith summed up: "In the great revolution . . . of which his [Smith's] rejection was the first wave, Connecticut abdicated her Christian standing. The ancient spirit which had shaped her institutions, and linked her, in her corporate capacity to the throne of the Almighty for almost two hundred years was then expelled . . ." William W. Andrews, "Eulogy," in John Cotton Smith, *The Correspondence and Miscellanies of John Cotton Smith* (Hartford: Harper & Brothers, 1847), 40.
[2] Knight defeated six-term governor William Jones by seventy one votes, out of 7,832 cast.
[3] There Governor John Brooks was comfortably reelected in April.
[4] Dwight, *Theology Explained*, 41.

a decade into his long presidency, college officials would applaud him for "liberating the American college from an excessive religious orientation."[1] In hidebound Connecticut, the intellectual revolution was now complete.

[1]Quoted in Kelley, *Yale*, 164.

Chapter 4. Who Hurls the Thunderbolt Now?

If we begin with certainties, we will end in doubts,
but if we begin with doubts and bear them patiently,
we may end in certainty.

— Francis Bacon

The scientific and religious quests for truth had coexisted since the Enlightenment. But efforts to reconcile them grew more difficult, as closer observation of natural phenomena attributed their causes to natural forces, not the hand of God. Because its effects were so widespread, the disastrous weather in 1816 strengthened the scientific viewpoint, reinforcing what had been learned from previous disasters, such as the earthquake which struck Boston on the early morning of 18 November 1755.

He was abruptly awakened, an hour or so after he had finally dozed off, by a sudden shaking of the house. He realized at once it was an earthquake. He recalled how that had felt when it had happened before, when he was just a boy of thirteen. Habituated to observe the natural world, he wanted to get up, cross his bedroom to the window, and see what the sky looked like. But he was afraid of being thrown to the floor. After lying still a moment, he sat up, lit a candle and checked his clock, noting the exact time when the shaking had started. It was 4:15 a.m. As he jotted this down, the house started "rocking and cracking" again — this time more wildly, the window panes, doors, and chairs rattling as if some angry subterranean spirit was jerking them back and forth. Fearing that the roof might cave in at any minute, he made his way gingerly down the stairway, unbolted the front door, and stepped outside. Alone in the darkness, he could not hear anything and surmised that the quake was over. But in that instant the ground under his feet began rumbling again. He hurried back inside and up the stairs to check the clock. He estimated the aftershock lasted for between seven and eight seconds.

Another jolt occurred an hour and ten minutes later, but this one was scarcely noticeable. So John Winthrop — Hollis Professor of Mathematics and Natural and Experimental Philosophy at Harvard, the most respected American scientist of his day, and great-great-grandson of a founder of the Massachusetts Bay Colony — dressed and went outside again to assess the damage. He walked up and down debris-littered streets of Boston, taking notes. More than one hundred chimneys had toppled over. Down by the harbor, cobblestones were strewn with bricks dislodged from chimneys and from the gable ends of houses. In church towers the hands of clocks stood still. Several wooden buildings were torn apart. A large wood distiller's cask had been wrenched open, and one of its staves hurled several feet into a nearby fence. Winthrop scanned the sky again, this time looking for any hint of the "glimmering light" some claimed preceded an earthquake, foreshadowing God's intent, but there was none. He remembered that — before he had gone to bed — the night sky had been clear under a nearly full moon, but with no hint of anything unusual in the offing. Could one still believe there was a message to be gleaned from a disaster like this?[1]

As a man of science, the forty-one-year-old Winthrop had been trained to gather information, not to speculate on ultimate causes. This was better left to the theologians, he had been taught. Man's power to understand the natural world was limited. Only God could fathom the marvelously intricate vision which had created all the creatures of the earth and the forces which operated on them, causing a sparrow to plunge from the sky or the earth to tremble and spew forth fire. Only God fully comprehended the natural laws which governed the universe and had sustained life since the time of Creation. Only God possessed the power to violate those laws if it so pleased Him, in His righteous indignation at the sinful ways of the human race He had made in His own image. But this earthquake on the morning of 18 November 1755 made Winthrop wonder if this explanation was valid. His observations led him to other speculations. The bricks had all fallen in the same direction — from the northwest to the southeast. So had a key on one of his shelves. Some bricks had flown sideways as much as thirty feet. It appeared that the earthquake had not reared up in one spot, but rippled across the city like a tidal wave. This observation tended to cast doubt on the belief that quakes were expressions of God's anger at a particular group of people, gathered at one place. So did

---

[1] John Winthrop, untitled article, *Boston Gazette*, 24 November 1755.

the fact — as Winthrop would later learn — that this quake had caused destruction over a wide area: centered off Cape Ann, it had been felt as far away as Nova Scotia and South Carolina. In New Hampshire, a zigzagging, one thousand-foot-long crack in the earth had opened up.[1]

The sheer magnitude of this seismic event was stunning: in recorded history, no previous earthquake of this size had ever struck New England.[2] Never before had damage been so widespread. According to the eyewitness account of one Bostonian, people leaped out of their beds at the first swaying of their houses and ran through the streets screaming "with the apprehension of its being the day of judgment."[3] In nearby Braintree, John Adams, fresh out of Harvard, lay motionless for four long minutes in bed at his father's house as it "seemed to rock and reel and crack," fearing it was about to collapse.[4] After the shaking subsided, residents of Salem, New Haven, and Boston stumbled through the streets looking as if they had just seen a ghost. Aftershocks persisted for over a month. People were in a state of shock. The ground beneath their feet no longer felt solid. They exchanged nervous glances which posed the questions none knew how to answer: *Why did this happen? What does it all mean?* Coming only a few years after powerful quakes had shaken Europe —London in 1750 and, most recently, Portugal — and after British losses in the French and Indian War had risen to an appalling level, many wondered what could have brought on this string of seemingly connected disasters. As devout believers, they sought answers in prayer. Many were moved to make amends with God, as the hour of judgment appeared at hand.

Meanwhile, John Winthrop had been conducting his own inquiry.[5] His analysis of the quake led him to different conclusions. He decided he should make these public. Winthrop felt compelled to do this because Boston's clergy were now proclaiming that this quake was a heavenly rebuke for sinful behavior.[6] The Reverend Jonathan Mayhew, of Boston's Old West

---

[1] Bryce Walker, *Earthquake* (Alexandria VA: Time-Life Books, 1982), 49.

[2] It has been estimated that the quake measured 5.8 on the Richter scale. No seismic event in New England since has reached this level.

[3] Quoted from an anonymous diary in Ballard C. Campbell, *Disasters, Accidents, and Crises in American History: A Reference Guide to the Nation's Most Catastrophic Events* (New York: Facts on File, 2008), 1767.

[4] Quoted from Adams's diary in I. Bernard Cohen, *Science and the Founding Fathers: Science in the Political Thought of Thomas Jefferson, Benjamin Franklin, John Adams, and James Madison* (New York: Norton, 1995), 213.

[5] In this endeavor he broke ranks with his illustrious, eponymous ancestor, who had scoffed at the notion of any events being random, and described them as a "connected chain, and required that they look within their hearts to find the cause of God's rebukes." Quoted in Virginia Anderson, *New England's Generation: The Great Migration and the Formation of Society and Culture in the Seventeenth Century* (Cambridge: Cambridge University Press, 1991), 195.

[6] While all New England ministers agreed that the earthquake was caused by

Church, had opined that both this earthly upheaval and the imminent French invasion of the British colonies were God's punishment for errant ways.[1] The lieutenant governor of Massachusetts, Spencer Phips, declared a day of fasting and atonement, deeming this necessary since it had "pleased Almighty God, in a most awful and surprising manner to manifest his righteous anger against the provoking sins of men by terrible and destructive earthquakes and inundations in divers parts of Europe and by a late severe shock of an earthquake on this continent and in this province in particular . . . ."[2]

One essay in particular — *Earthquakes, the Works of God and Tokens of His Just Displeasure* — riled Winthrop. Written by the Reverend Thomas Prince following a tremendous New England quake in 1727, this piece had been hastily republished a week after this latest one, to chastise the populace. The clergyman's argument for divine causation flew in the face of all that Winthrop had witnessed, and so he could not let it go unchallenged. He knew Prince was flatly wrong: this quake had struck too wide an area to claim that Boston was being singled out for reproach. Winthrop could not accept this warmed-over assertion that God intervened in the natural world to display His anger. It was his view that the Deity, having set the universe in motion in accordance with immutable laws, remained aloof from its daily operations, much as a man who winds a clock and then lets it run for days without touching it. Winthrop was determined to present his scientifically enlightened theory in the interests of improving understanding of how physical processes actually worked. Therefore he decided to change the subject of a talk he had agreed to give on November twenty-sixth in Harvard's chapel to this pressing matter of earthquakes.[3]

As he rose to speak before a hushed, expectant throng of students on that appointed day, Winthrop could not divine how they would react. Buildings on the campus had been shaken by the quake, and so the students' curiosity in this subject was not at all idle. Winthrop had been giving lectures on electricity during the past decade, but that did not mean that all those in the audience accepted the teachings of science.[4] Most Bostonians, in fact,

God, they differed on the "divine mechanism and motivation" responsible. See "Two Boston Puritans on God, Earthquakes, Electricity, and Faith: 1755-1756," National Humanities Center Resource Toolbox: "Becoming American: The British Atlantic Colonies, 1690-1763." http://nationalhumanitiescenter.org/pds/becomingamer/ideas/text1/godlightningrods.pdf.
[1] Rev. Jonathan Mayhew, *Practical Discourses Delivered on Occasion of the Earthquakes in November 1755* (Boston: Richard Draper, 1760). Cited in Campbell, *Disasters*, 1767.
[2] Sidney Perley, *Historic Storms of New England* (Salem MA: Salem Press, 1891), 62-3.
[3] Most of the material he presented during this lecture was contained in Winthrop's 24 November article in the *Boston Gazette*.
[4] Philip Dray, *Stealing God's Thunder: Benjamin Franklin's Lightning Rod and the Invention of America* (New York: Random House, 2005), 107.

sided with the clergy.[1] He could only hope that his findings would engage their young minds. An earthquake, he told his listeners, was an "agitation or shaking of some considerable part of the earth, and that by natural causes." He emphasized the word "natural," to make clear the direction his remarks were going to take. Winthrop then summarized what had just befallen New England, creating so much havoc and causing so many people to fear for their lives. He told them of his own sensory experiences — the rumbling sound before the shaking had begun, the vibrations of his house, the clear black sky over Boston, bricks on his hearth undulating under his feet during an aftershock. Then Winthrop reviewed the explanations offered by theologians and quoted biblical descriptions of boulders being heaved up from under the earth, the seas frothing with gigantic waves, ships being tossed around like toys, people scattering like ants after a boot has crushed their mound.

He countered these accounts with his own theory of how earthquakes arose from deep within the earth. Harkening back to Aristotle, he postulated that the earth was riddled with "large holes, pits, and caverns," where new materials such as iron and sulfur were formed, under conditions of great heat. Because these materials were combustible, their attraction could result in violent explosions, and the ensuing expanding gases cause the ground to "heave up." Thus earthquakes had purely physical causes. But this did not rule out the hand of God. Destructive quakes could "justly be regarded as the tokens of an incensed DEITY," Winthrop said. They could also serve a beneficial purpose. The release of gases he likened to the pores of the skin opening up to release toxins and purge the body: it made soils ready for seeding and thus helped to sustain human life. Though he therefore did not doubt that there was an ultimately moral reason for such phenomena, Winthrop did not attempt to explain this. He was only capable of discussing *natural* philosophy. Only God knew how good could emerge from "the greatest evils."[2]

A few days later the Reverend Prince published a lengthy rebuttal of this Harvard lecture in a letter in a Boston newspaper, which slyly tried to turn the tables on the professor. Here Prince fired another salvo at scientific "progress" by blaming lightning rods — invented by Benjamin Franklin in 1752 — for causing so much damage to Boston: these supposedly protective objects had drawn electricity down from the sky, igniting underground explosions.[3] Prince mocked the idea that these iron rods could thwart the will

[1] John Adams so observed in notes he made at that time. Cohen, *Science and the Founding Fathers*, 214.
[2] John Winthrop, *A Lecture on Earthquakes, Read in the Chapel of Harvard College, in Cambridge, N.E., November 26th, 1755, on Occasion of the Great Earthquake Which Shook New England the Week Before* (Boston: Edes and Gill, 1755), 5, 26, 29, 31.
[3] At that time, the Massachusetts capital sported more protective iron rods than other U.S. cities. Homer W. Smith, *Man and His Gods* (Boston: Little,

of God by steering lightning bolts away from their preordained destination. Earthquakes, he would subsequently declare, were the "Resentments of Almighty GOD" against human sins.[1]

A highly-educated (Harvard) man who read the Bible in its original languages, the long-time pastor of Boston's historic Old South Church was unshakably wedded to Calvinist teachings. Foremost among these was the Almighty's protection of those who honored and submitted to Him.[2] A decade earlier, Prince had perceived this benevolence in the victory of New England colonial forces over the French at Louisbourg, on Nova Scotia's rocky Cape Breton: an unusually mild winter had enabled their fleet to sail the treacherous North Atlantic without losing a single ship so that they could land and lay successful siege to this supposedly impregnable fortress.[3] Prince supported empirical investigations like Winthrop's, confident they would confirm that science and religion were not incompatible.[4] Prince's conviction was shared by many theologians of his day and well into the nineteenth century, as well as by natural philosophers.

The reverend and the professor continued their lively debate in the pages of the *Boston Gazette* as well as in a series of pamphlets.[5] Winthrop questioned Prince's grasp of science: if lightning bolts caused earthquakes, how could the latter be felt over such a wide area? He also attacked the

---

Brown, 1952), 295. Residents of Boston had embraced Franklin's invention, having experienced the ravages of fire just a few years before, in 1747. Dray, *Stealing God's Thunder*, 105.

[1] Prince, Thomas, "Boston, December 29," *Boston Gazette*, 29 December 1755.

[2] *Catalogue of the Collection of Books and Manuscripts Which Formerly Belong to The Reverend Thomas Prince* (Boston: Alfred Mudge and Son, 1870), vi-vii.

[3] Thomas Prince, *Extraordinary Events, the Doings of God, and Marvellous in Pious Eyes*, 3rd ed. (Boston/New York: J. Lewis, 1745), 21, 28. Prince's belief in the efficacy of prayer was reinforced a year later when his imprecations that strong winds arise to prevent French ships from attacking Massachusetts were followed by a storm which wrecked the enemy fleet. John S. Friedman, *Out of the Blue: A History of Lightning: Science, Superstition, and Amazing Stories of Survival* (New York: Random House, 2008), 83.

[4] This view was shared by earlier Puritan leaders such as Increase Mather, who was a friend of the British chemist, natural philosopher, and inventor Robert Boyle. The essays his son, Cotton, published in 1721 under the title of *The Christian Philosopher* covered subjects such as the sun, comets, the planets, lightning, and the earth. The younger Mather asserted that his opinions reflected the "learning of the times." Abijah P. Marvin, *The Life and Times of Cotton Mather* (Boston, Chicago: Congregational Sunday-School and Published Society, 1892), 456. Cotton Mather was one of first Americans to become a member of the prestigious Royal Society of London.

[5] Winthrop published an appendix to his lecture on earthquakes, which was followed by a letter of Prince's to the *Boston Gazette* on 26 January 1756. Winthrop responded with a pamphlet entitled *A Letter to the Publishers of the Boston Gazette*. For the chronology of their exchange, see Christopher Grasso, *A Speaking Aristocracy: Transforming Public Discourse in Eighteenth-Century Connecticut* (Chapel Hill: University of North Carolina Press, 1999), 215, n. 40.

---

minister's "pathetic exclamation" that persons who used rods to protect their buildings were trying to thwart the "mighty hand of God." Winthrop accused local clergy of manipulating fears about disasters like this in order to keep the faithful from straying.[1] At stake here was the nature of God's relationship to His creation: did He cause earthquakes, floods, and volcanic eruptions, or were they simply consequences of natural laws and thus unrelated to human behavior, without moral significance?

This question became more pressing when news arrived, in late December 1755, of a devastating quake in Lisbon.[2] On the morning of 1 November, a massive undersea explosion, followed by three tsunami waves over twenty feet high, had battered the city, toppling most of its buildings and spreading flames which incinerated the debris and killed an estimated sixty thousand persons.[3] Fires continued to burn for nearly a week, gutting most of central Lisbon and environs, turning them into a "charred desert of smoking ruins," littered by dead bodies.[4] Early accounts stressed that the "Hand of God" was responsible.[5] The Reverend Prince attributed this earthquake to God's resentment over the "wickedness" of man.[6] It seemed there had to be some supernatural connection between these two seismic events, coming so close together on either side of the Atlantic. But Winthrop thought this was merely coincidence. His disagreements with Prince mirrored the wider debate in European capitals, where eminent philosophers, clergy, and average citizens tried to make sense of the destruction of the Portuguese capital. There, too, men of the cloth declared it the act of a God incensed by neglect of His commandments. However, this argument strained credulity, since the quake had occurred on All Saints' Day — the judgment day of the Inquisition —

---

[1] Quoted in I. Bernard Cohen, *Benjamin Franklin's Science* (Cambridge: Harvard University Press, 1990), 153.
[2] See, for example, "Extract of a Letter from Cadiz, Nov. 4 1755," *Boston Gazette*, 22 December 1755. This account had reached Cape Ann via a ship which had sailed from the Portuguese capital shortly after the quake struck. Cf. "Extract of a Letter dated at Belem, near Lisbon, Nov. 6. 1755," *Boston Evening Post*, 29 December 1755.
[3] T.D. Kendrick, *The Lisbon Earthquake* (New York, Philadelphia: J. Lippincott, 1905), 56.
[4] James Chase, "The Great Earthquake at Lisbon," *The World's Great Events, In Five Volumes: A History of the World from Ancient to Modern Times, B.C. 4004 to A.D. 1903*, vol. 4, *Colonization and Discovery, A.D. 1704 to A.D. 1830*, ed. Esther Singleton (New York: P.F. Collier and Son, 1904), 1613. Chase, then twenty six, was an eyewitness to the quake.
[5] "Extract of a Letter from Cadiz," *Boston Gazette*, 22 December 1755. Chase reported that priests in Lisbon told survivors the quake was "a judgment on them for their wickedness." Chase, "Earthquake," 1616.
[6] Anonymous article, "Boston, December 29," *Boston Gazette*, 29 December 1755.

when the churches of Lisbon had been packed with devout Catholics.[1] Some forty churches had collapsed upon these worshippers, and nunneries had also been reduced to rubble. Many of the faithful had burned to death inside churches where they had sought refuge.[2] Why would a loving Deity allow so much death and suffering to rain down on this loyal flock?[3]

Gottfried Leibniz, who had invented calculus and then declared it was evidence of God's benign plan for the universe, published three papers reaffirming that even an earthquake-prone planet was still "the best of all possible worlds." His sanguinity was shared by contemporaries such as Alexander Pope. From Geneva, Voltaire dashed off a poem ridiculing this "naïve" credo.[4] He subsequently demolished Enlightenment optimism in his devastating satire, *Candide*. For his part, Jean-Jacques Rousseau interpreted the Lisbon quake as proof of the "essential evil of civilization." A confused, six-year-old Goethe reflected upon this horrible event and found it hard to believe that it manifested the hand of a "fatherly character."[5] His later investigations of phenomena such as color and plant structure, as well as the weather, would express a religious skepticism engendered, in part, by this childhood experience. [6]

The tragic loss of life in this 1755 earthquake spurred the application of

---

[1] Karl Fuchs, "The Great Earthquakes of Lisbon 1755 and Aceh 2004 Shook the World: Seismologists' Societal Responsibility," in *The 1755 Lisbon Earthquake: Revisited*, ed. Luiz A. Mendes-Victor, Carlos Sousa Oliveira, Joao Azevedo, and Antonio Ribeiro (New York: Springer, 2005), 46.

[2] Kendrick, *Lisbon Earthquake*, 59-60.

[3] Historically, churches were struck far more frequently by lightning than other structures because their steeples rose so high. Cotton Mather conceded as much when he wrote: "Our meeting houses and our ministers' houses have had a singular share in the strokes of thunder." Quoted in Perley, *Historic Storms*, 83.

[4] In his "Essay on Man," Pope had written that "whatever is, is right." Voltaire's poetic retort, "An Inquiry into the Maxim, 'Whatever Is, Is Right,'" contains this stanza: *All's right, you answer, the eternal cause / Rules not by partial, but by general laws. / Say what advantage can result to all, / From wretched Lisbon's lamentable fall? / Are you then sure, the power which could create / The universe and fix the laws of fate, / could not have found for man a proper place, / But earthquakes must destroy the human race?* In Voltaire's preface to his poem on the Lisbon earthquake, he stated: "If a man is devoured by wild beasts, causes the well-being of those beasts, and contributes to the order of the universe; if the misfortunes of individuals are only the consequence of this general and necessary order, we are nothing more than wheels which serve to keep the great machine in motion; we are not more precious in the eyes of God, than the animals by whom we are devoured." Voltaire, *Candide and Other Writings*, ed. Haskell M. Block (New York: Modern Library, 1956), 194.

[5] Johann Wolfgang von Goethe, *Goethe's Autobiography*, part 1, *Goethe's Boyhood, 1749-1764*, trans. John Oxenford (London: William Clowes and Sons, 1888), 19.

[6] Goethe helped to develop a "weather ball" barometer and began using it in the early 1820s. By 1822 he was comparing barometric pressures from places around the world, including Boston, London, Vienna, and Weimar.

---

scientific reasoning and methodology to understand other natural events — from comets streaking across the night sky to the movement of the wind.[1] Emmanuel Kant, a proponent of the "Great Separation" between the natural and supernatural worlds, compiled a record of observations during the Lisbon quake. John Winthrop's notations in Boston also added significantly to knowledge about these sudden, violent events. For this pioneering work, he would come to be known as the "father of seismology."[2] In the following decades, European investigators developed other theories about what caused quakes, with the hope that scientific insights could contribute to the well-being of society.[3]

Lightning had become a focus of inquiry after Franklin had demonstrated in 1752 that it could draw electrical energy down out of the sky.[4] To some it seemed that Prince may have been correct in blaming lightning rods for attracting large bolts and thus causing underground upheavals. However, if manmade devices could cause such a discharge of energy, what then was God's role in these events? If iron rods proved more effective in protecting churches from fire than prayers, the sprinkling of holy water, or the ringing of bells, did it still make sense to count on these practices?[5] More and more educated, knowledgeable persons began to have doubts. The deaths of so many believers in Lisbon only seemed explicable as an act of some natural force. To think otherwise was to worship a Deity who punished without purpose or just cause. Few were prepared to accept this notion. Instead, they tended to agree with philosophers like Locke, Hume, Kant, and Hobbes, who saw the Almighty as removed from the daily workings of the world. This

---

[1] Franklin's role in advancing the scientific method can scarcely be exaggerated. In the words of one historian of science and politics in early America, Franklin's invention of the lightning rod made clear that "pure and disinterested scientific research of a fundamental kind would ultimately lead to inventions of use to humankind." Cohen, *Science and the Founding Fathers*, 147. His seminal studies of electricity are his most well-known experiments, but Franklin also made important findings about the weather, plotting — for example — the course of the Gulf Stream during a westward voyage across the Atlantic in 1775. See Walter Isaacson, *Benjamin Franklin: An American Life* (New York: Simon and Schuster, 2003), 290. He would resume this study during his last crossing, in 1785. Dray, *Stealing God's Thunder*, 175-6.

[2] See, for example, Ken Fitzgerald, *Weathervanes and Whirligigs* (New York: Crown, 1967), 123.

[3] Fuchs, "Great Earthquakes," 44.

[4] John Winthrop's experiments with electricity, for example, owed much to Franklin's discovery. See letter of Winthrop to Franklin, 29 September 1762, *Experiments and Observations on Electricity Made at Philadelphia in America, by Benjamin Franklin*, 4th ed. (London: D. Henry, 1769), 434.

[5] The ringing of bells had continued as a means of deterring lightning strikes on Protestant churches up until the end of the seventeenth century. Their sounds were believed to scare away the demons which hurled down lightning bolts. Andrew Dickson White, *A History of the Warfare of Science with Theology in Christendom*, vol. 1 (New York: Appleton, 1922), 342.

growing acceptance of the Great Separation marked a sea change in Western thinking. Up until now, study of the natural world had been conducted to find evidence of the Almighty's grand design. For example, earlier in the eighteenth century, the English philosopher David Hartley had written that all pursuit of truth should be "entered upon with a view to the glory of God, and the good of mankind . . . ."[1] By eliminating this rationale, "the whole tremendous fabric of theological meteorology reared by the fathers, the popes, the mediaeval doctors, and the long line of great theologicans [sic], Catholic and Protestant, collapsed; the 'Prince of the Power of the Air' [the Devil] tumbled from his seat; the great doctrine which had so afflicted the earth was prostrated forever."[2]

In fact, the transformation in outlook was neither so dramatic nor so abrupt. Doubts about supernatural casualty increased, but faith persisted. This was certainly the case in the United States, where religious revivals in the late eighteenth and early nineteenth centuries stressed salvation through conversion and unquestioning faith. At the same time, more lightning rods were installed. Franklin claimed that they were commonplace by 1772 — far more prevalent than in France a dozen years later.[3] Still, it is probably more accurate to say that, at the time of the Revolution, most Americans still agreed with Thomas Prince and his fellow Calvinists that earthquakes were determined by God and thus opposed the use of lightning rods, for these dared to avert the expression of divine "displeasure."[4] Skepticism about their efficacy — and religious propriety — persisted, even though some congregations quietly placed rods around their churches.[5]

The American Revolution was spurred by a desire for political rights.

---

[1] Quoted in Joseph Priestley, preface, *The History and Present State of Electricity with Original Experiments*, vol. 1 (London: C. Bathurst and T. Lowndes, 1755), xxvi.

[2] White, *Warfare of Science*, 364. A scientist, writing in the 1950s about the supernatural in human affairs, noted that "the last remnants of meteorological demonology crumpled down" after Franklin had demonstrated the power of the lightning rod. Smith, *Man and His Gods*, 295. But Smith also noted that many New Englanders continued to regard Franklin as the "archinfidel." Belief that the Devil was responsible for lightning, hailstorms, thunder, and droughts grew out of pagan beliefs that "evil gods" caused such events.

[3] Dray, *Stealing God's Thunder*, 153. Franklin wrote that, as early as 1752, rods had "become so common, that Numbers of them appear on private Houses in every Street of the principal Towns, besides those on Churches, Publick Buildings, Magazines of Powder, and Gentlemens [sic] Seats in the Country." Letter of Franklin to Horace Benedict de Saussure, 8 October 1772, *The Papers of Benjamin Franklin*, vol. 19, *January 1, 1772, through December 31, 1772*, digital edition, Packard Humanities Institute, http://franklinpapers.org/franklin/framedVolumes.jsp?vol=19&page=324a.

[4] White, *History of Warfare with Science*, 366.

[5] Hermann Melville portrayed one such skeptic — and an unscrupulous salesman of these devices — in his 1854 short story "The Lightning-Rod Man."

Henceforth, figures of authority could not count on maintaining power through meek subordination. This insurrectional mood carried over to organized religion. Revolutionary leaders like Thomas Paine (in his *Age of Reason*) openly questioned the tenets of orthodox Christianity. Thomas Jefferson and other Deists veered away from belief in an omnipresent Deity. They professed that, if the Almighty did not determine the course of human events, then it was up to His human creatures to exercise their God-given powers of inquiry, "common sense," and cogitation to steer a course leading to salvation and eternal life. A nascent Unitarian-Universalist church urged its members to trust their own reason to interpret the Bible and discover God's truth. Christians should only believe that which they could understand.[1] So preached Caleb Rich, the powerful voice of Universalism in rural western Massachusetts, spreading a credo of self-reliance and self-determination while disavowing any idea of preordained eternal punishment in hell.[2] This new humanistic and democratic-minded denomination, growing out of Pietistic and Anabaptist movements in Europe, had first taken hold in the Mid-Atlantic states and then spread north into New England.[3] Boston had its first Universalist congregation in 1787.[4] Soon Universalism extended westward, into New York and Pennsylvania, as an unhappy Thomas Robbins discovered on a journey to Ohio.[5]

Because of its methodology (open inquiry) and spiritual goal (knowledge based on truth), Unitarian-Universalism was a natural fit for the age of political democracy ushered in by Jefferson's election in 1800. To many prominent Democratic-Republicans, exploring the physical world and learning how it worked was a central responsibility of citizenship — producing knowledge which sustained the power of the people.[6] Jefferson,

---

[1] Thus some critics of Universalism expressed this tenet. See, for example, Daniel Hascall, *Caution against False Philosophy: A Sermon* (Hamilton, NY: Johnson and Sons, 1817), 8. The Unitarian leader William Ellery Channing put it in these words: ". . . the existence and veracity of God, and the divine original of Christianity, are conclusions of reason, and must stand or fall with it." Channing, "Discourse at the Ordination of the Rev. Jared Sparks, Baltimore, 1819," in Channing, *The Works of William E. Channing, D.D.*, vol. 3, 11th ed. (Boston: George G. Channing, 1849), 66.

[2] Hatch, *Democratization of American Christianity*, 40-1.

[3] Joseph Priestley, the eminent British chemist, natural philosopher, and religious dissenter, had brought his Unitarian faith to Philadelphia when he fled there to escape persecution in 1791.

[4] Berryman, *From Wilderness to Wasteland*, 97.

[5] Thomas Robbins, *Diary of Thomas Robbins, D.D.*, vol. 1, *1796-1854* (Boston: Beacon, 1884), entry for 2 October 1803, 207.

[6] The wide dissemination of information was a major objective of Democratic-Republicans early in the nineteenth century. An almanac for the year 1816 noted with approval the benefits to human well-being which had resulted from this trend: "The diffusion of information on every subject of utility is essential to the preservation of our republican forms of government, as

for one, believed that political principles were based upon laws, just as those in the physical world were.[1] Better understanding of natural processes and the practical application of this knowledge would benefit all of humanity. For all these reasons, Jefferson was a strong proponent of science: his enthusiasm ranged from examining dinosaur bones at his executive mansion, to backing the creation of West Point in order to promulgate military science, to sending Lewis and Clark west with instructions to collect flora and fauna samples along the way.[2]

Because of the weather's importance to an agriculture-based American economy, forecasting it was critically important. But, by the start of the nineteenth century, little progress had been made. As was noted earlier, the keeping of meteorological records, by academics and interested amateurs, had started back in the 1790s, thanks to the refined calibration of Fahrenheit's thermometer. Still, more accurate measurement had not led to speculations about underlying causes. This prevailing reticence did not deter an inquisitive Thomas Jefferson from analyzing the weather. As a Virginia plantation owner, he had a pressing need to know what the next season would bring. In the early 1780s he observed that recent winters at Monticello had brought less snow and deduced that the earth was warming.[3] Benjamin Franklin similarly had a lifelong fascination with the weather. As a twenty-year-old, sailing to Philadelphia from London in 1725, he had made detailed notations about the rainbows visible on board.[4] In 1743, while serving as Philadelphia's postmaster, Franklin had asked colleagues in several other colonies to help him track a hurricane headed toward New England from the West Indies. Riding on horseback into a whirlwind in Maryland, the arch-empiricist lashed at it vainly with a whip — to better understand its movement.[5] In 1783, he published a paper theorizing (correctly) that ash

---

well as to free them from the unfavorable imputations which have but too frequently and sometimes with too much justice been attached to their management." *An Agricultural and Economical Almanack, for the Year of Our Lord 1816. Calculated for the Meridian of New Haven — Lat. 41, 18'* (New Haven: Society for Promoting Agriculture in the State of Connecticut, 1815). [not paginated].

[1] Garry Wills has argued that the Declaration of Independence was heavily influenced by Newtonian science. See Cohen, *Science and the Founding Fathers*, 110-11. Benjamin Franklin had proposed a change in Jefferson's original draft dealing with fundamental truths, replacing the adjectives "sacred" and "undeniable" with "self-evident." Isaacson, *Franklin*, 312.

[2] George Brown Goode, "The Origins of the National Scientific and Educational Institutions of the United States," *Papers of the American Historical Association* 4:2 (20 April 1890): 20.

[3] Bernard Mergen, *Snow in America* (Washington, DC: Smithsonian Institution Press, 1997), 1.

[4] Dray, *Stealing God's Thunder*, 29.

[5] Letter of Franklin to Peter Collinson, 25 August 1775. Quoted in Isaacson, *Franklin*, 166.

from erupting volcanoes could lower temperatures.[1]

Jefferson and Franklin belonged to an estimable fraternity of "amateur" scientists, whose ranks included such pioneering figures as William Herschel, his sister Caroline, Humphry Davy, and Joseph Banks. Without institutional backing and with limited resources, they managed to add much knowledge in their fields.[2] In fact, these men (and a few women) did more to increase the understanding of physical laws and processes than their contemporaries at colleges and universities. Unfortunately, their eclectic interests and lack of rigorous training limited their investigations. Franklin may have had an illuminating insight into how particles in the air could cool earth temperatures, but he had no means — or inclination — to test this theory.

Because of their complexity and rarity, phenomena like earthquakes could not readily be studied. Rainstorms and droughts, spells of intense heat and bone-chilling cold, appeared to arise unpredictably, without any discernible causes. These weather events were not like natural events which occurred with regularity and thereby confirmed the Newtonian notion of a world running like clockwork. Throughout most of human history, much about the weather had defied close observation because no instruments existed which could precisely gauge its fluctuations. Until the invention of the thermometer and the barometer, in the seventeenth century, there had been no reliable means of tracing trends over long periods of time. For two millennia before then, changes in the weather had been largely a matter of conjecture, derived from the theories of Aristotle.[3] Added to these was the "wisdom" accumulated over centuries from folklore, superstition, magic, and astrology. Belief in such unscientific explanations, coupled with religious convictions, prevented the development of a more rational approach to predicting rainfall, storms, and other events.[4]

With the advent of these new instruments, fact-based theories were advanced in the second half of the eighteenth century, in the wake of Franklin's experiments with electricity. Hopes arose of turning meteorology into a genuine science. George Adams, an English purveyor of weather devices (and, hence, not an impartial commentator) wrote in 1790 that the

---

[1] He based this theory on the severely cold winter which followed an Icelandic eruption in 1783 and 1784. H.E. Landsberg and J.M. Albert, "The Summer of 1816 and Volcanism," *Weatherwise* 27:2 (April 1974): 64.

[2] For an excellent depiction of these scientists and their contributions, see Richard Holmes, *The Age of Wonder: How the Romantic Generation Discovered the Beauty and Terror of Science* (New York: Pantheon, 2008).

[3] For example, Aristotle had surmised that meteors were caused by "exhalations" between the earth and the sun. Vladimir Jankovic, *Reading the Skies: A Cultural History of English Weather, 1650-1820* (Chicago: University of Chicago Press, 2001), 17.

[4] Cox, *Storm Watchers*, 4.

time had finally arrived when "plain and sound observations" could yield reliable theories, leading to more accurate predictions.[1] But this did not work out as Adams had hoped. Thermometer readings did not reveal any telltale pattern or underlying laws.[2] Sometimes the insights gleaned from limited experimentation did prove enlightening, as when the self-taught English chemist and meteorologist John Dalton speculated — correctly — that trade winds were propelled by the earth's rotation and differences in atmospheric temperature.[3] But many theories were wildly misleading. As one present-day historian has put it, "Meteorology had to wait for other sciences to define its basic principles and forces: relations of energy and mass, gravity and friction, thermodynamics, and the behavior of gases."[4]

Thus, at the start of the nineteenth century, much remained frustratingly unclear about the weather. Londoners, for instance, complained about the shroud of fog which so often enveloped their neighborhoods, but scientists had no idea how to reduce or eliminate it.[5] What was missing was a grasp of the big picture. Isolated from each other, unable to communicate in a timely fashion over great distances, pre-modern humans rarely knew what was taking place elsewhere on the planet. Thus, all weather was local. This parochialism reinforced pagan and Christian beliefs that extraordinary weather events, as well as volcanic eruptions, shooting stars, and earthquakes, had cosmic causes but a targeted human focus. Because these events were frequently awe-inspiring in their power, it was logical to ascribe them to an omnipotent deity. Mundane occurrences like an afternoon shower or light breeze seemed less portentous and thus more likely to have purely physical causes. But even these were not readily apparent. Measurements of temperatures and snowfall in places like New Haven and Williamstown did not explain why one day was warmer than the next, or why a village received more rain one year than it had the one before.

With the growth of newspapers and transatlantic travel, weather watchers came to realize that climate patterns extended over vast territories. To understand what the weather was apt to be like tomorrow, one had to know what it had been a few days or weeks before in distant locations. The

---

[1] George Adams, *A Short Dissertation on the Barometer, Thermometer, and other Meteorological Instruments* (London: R. Hindmarsh, 1790), iv-vii.

[2] John Henry Belville, *A Manual of the Thermometer, Containing Its History and Use as a Meteorological Instrument* (London: Manual, 1850), 12.

[3] This theory had first been proposed by another amateur meteorologist, George Hadley, in 1735, but Dalton had been unaware of Hadley's paper describing it.

[4] Cox, *Storm Watchers*, 4.

[5] Jankovic. *Watching the Skies*, 14. It was not until the end of the nineteenth century that proposals were advanced to reduce fog by switching from coal fires to gas. See, for example, "The Exorcism of the Smoke Fiend; Or, How to Get Rid of the Plague of Fog," *The Review of Reviews* 5:27 (March 1892): 298.

first attempt to compare simultaneous readings at different places was made by two future American presidents — Thomas Jefferson and James Madison — in the late 1770s. They shared notations about the temperatures and other weather conditions at Monticello and the College of William and Mary.[1] The next step in this direction was taken by the U.S. Army Medical Department, at the behest of the Surgeon-General, General James Tilton, in 1814. Hospital surgeons and other medical officials at Army posts throughout the country were ordered to keep records of the temperature, barometric pressure, level of moisture, rainfall, wind direction and strength, and other weather-related phenomena.[2] The purpose was to determine if any meteorological changes were harmful to human health. An outbreak of "bilious fever" during 1812 and 1813 had prompted this undertaking. Data were gathered from Eastport, Maine, to Prairie du Chien, Wisconsin — from Fort Niagara on Lake Ontario to New Orleans.[3] Over the next three decades, these records were collated and published in three volumes under the title of *Meteorological Registers*. But these reams of figures did not show how weather and health might be connected.

By the summer of 1816, the value of keeping records of local temperatures and other weather occurrences was still being debated. Natural philosophers saw gathering such data as a way of cataloguing the minute particulars of a universe they could not grasp in its entirety. Their faith restrained them from probing further. One such scientist was Chester Dewey, the son of a Sheffield, Massachusetts, farmer, whose unusual intellectual gifts and intense curiosity about the world around him gained him admission to Williams College in 1802. But, by his senior year, spirituality had become paramount in his life, and, feeling guilty for his previous "ingratitude" toward God, he dedicated himself to "nobler impulses than the promptings of natural sympathies." After considering a career as a Congregationalist minister, he returned to his alma mater as a tutor in the natural sciences. There Dewey championed new fields, such as geology. At the same time, he took pains to see that his students developed into good Christians. One year a junior asked that his class hold a prayer meeting, instead of the usual morning lecture. Dewey was delighted that so many students were prepared to denounce their godless ways and welcome the Almighty into their hearts: "Levity, recklessness were

[1] Mark W. Harrington, "History of the Weather Map," *Bulletin No. II, of the Weather Bureau*, (Washington D.C.: Department of Agriculture, 1895), 327.

[2] Because the United States was still at war in 1814, the gathering of this information was delayed until 1818, when Gen. Joseph Lovell had succeeded Tilton as Army Surgeon General. *Historic Climate Variability and Impacts in North America*, eds. Lesley Ann Dupigny-Giroux and Cary J. Mock (London, New York: Springer, 2009), 174.

[3] Joseph Lovell, "Meteorological Register for the Year 1820," in William H. Keating, *Narrative of an Expedition to the Source of St. Peter's River, Lake Winnepeek, Lake of the Woods, &c*, vol. 2 (London: George B. Whittaker, 1825), 140.

all gone — earnestness, honestness filled their souls — the depths of feeling stirred — tears flowing — prayers ascending."[1]

In 1811, the year after being appointed professor of mathematics and natural philosophy at Williams, Dewey began making meteorological observations. Three times a day he would consult a mercury thermometer outside his house and record his observations of temperature, rainfall, wind direction, and the nature of the weather. In addition to this daily routine, he made notes on sunspots, phases of the moon, and other celestial occurrences — all as a "means of understanding God."[2] Under Dewey's tutelage, a Meteorological Association was established at the college. His ground-breaking observations were published in the prestigious *American Journal of Science* and *North American Review*, and he became recognized as a leader in this emerging field of inquiry.

Parker Cleaveland's life followed a similar path. Born in Rowley, Massachusetts, in 1780, he was a precocious youth, shy and reserved. In 1795, at the age of sixteen, he entered Harvard, where he was quickly pigeonholed as a country bumpkin for wearing a felt hat considered hopelessly unfashionable by his more sophisticated classmates. Like Williams College, Harvard was then known more for "infidelity and misrule" than for observing the Sabbath. Cleaveland's preacher in Rowley had warned him about the dangers of associating with students there: "Examine religion for yourself; trust to no one else; then make a sacred vow not to depart from the religion of your fathers." Cleaveland took this advice to heart for the rest of his life: he never challenged authority or questioned orthodox religious and scientific principles. After graduating in 1799, he contemplated going into the law, but his parents urged him to become a minister instead. Dutifully he agreed to follow this path, but then was offered a post at Harvard, to teach mathematics and natural philosophy — in the same department where John Winthrop had taught until his death four years before.

Cleaveland accepted this invitation, but still worried that teaching these subjects was not the right thing for him to do. He kept asking himself, "Is it degrading to study nature, and the will and operations of nature's God?" A few months after returning to Harvard, he made a "confession of religion" in the nearby church where he had been baptized, hoping this step might end his inner turmoil. With Unitarian theologians like Henry Ware now in residence, and with Samuel Webber, a member of the same denomination, serving as president, Harvard was embracing a more open-minded approach to scientific inquiry.[3] Relieved that he did not have to choose between science

---

[1] Henry Fowler, *The American Pulpit: Sketches, Biographical and Descriptive, of Living American Preachers* (New York: Fairchild, 1856), 52-9.
[2] "Weather, Stars, and Living Nature."
[3] In 1805, Ware's appointment as the Hollis Professor of Divinity — the first

and his faith, Cleaveland soon relocated to Bowdoin, to continue teaching mathematics and natural philosophy. He gave the college's first courses in geology and chemistry and discovered a passion for the study of minerals. Even as his gained preeminence in this field, Cleaveland did not put forth any theories about the earth's history or origins, but stuck religiously to compiling facts. In the words of a memoirist, "His mind was practical and even materialistic in its turn, rather than speculative; clear in perception, rather than profound in insight . . . better skilled in the orderly arrangement of facts and the plain statement of laws, than in the deeper intuitions of higher generalizations of science; — a constitution of mind better adapted to the teaching, than to the discovery of truth, and to the teaching of the physical, than of the metaphysical sciences."

His daily routines were strictly followed. Cleaveland dedicated the same amount of time each day to preparing his classes, making observations, and taking his daily constitutional, deriving pleasure from so rigidly adhering to his schedule. The slightest deviation he could not tolerate, nor anyone who attempted to alter his plans: change was inherently bad. A proposal to move the date for Bowdoin's commencement precipitated this violent protest: "If that was unsettled, everything was deranged." Cleaveland's need for constancy carried over to his religious observances: abiding by the practice of his Puritan forebears, he set aside every Sabbath to reading religious tracts and every morning after breakfast studied passages from the Bible. He believed religion ought to serve as a counterbalance to an upstart, unbridled science, which otherwise would become "vainly puffed up" and go off in some radically "infidel direction." Whenever the choice was between the truth spelled out in Revelations and scientific speculations about the natural world, he would always side with the word of God.[1] Such religious caveats impeded the development of a meteorological science within the academy.

On top of this, most Americans already had other sources to turn to for predicting the weather. Chief among these were the almanacs which were printed annually in numerous editions — six in Franklin's Philadelphia alone — and sold in greater numbers than all other volumes. These tomes were typically infused with religious faith. In 1731, for example, the year before Franklin's *Poor Richard's Almanack* first appeared, Nathaniel Ames published one which advised that "the great GOD of Nature forewarns a sinful World of approaching Calamities, not only by Prophets, Apostles and Teachers, but also by the Elements and extraordinary Signs in the Heavens,

Unitarian to hold this prestigious post — ignited a firestorm of controversy at Harvard. Webber assumed the presidency the following year.
[1] Leonard Woods, *Address on the Life and Character of Parker Cleaveland, LL.D.* 2nd ed. (Brunswick ME: Joseph Griffin, 1860), 12-62.

Earth and Water."[1] Outlining what lay ahead in the year 1816, a New Haven almanac conceded the value of new information about how to ward off smallpox or a bolt of lightning, but affirmed that the hand of the Almighty still directed by "stated laws the minuter [sic] operations of nature" for the good of humanity. Readers could thus count on His promise "that day and night, cold and heat, summer and winter, seedtime and harvest, shall not fail until the consummation of all things."[2] In this religious environment, amateur scientists keeping records about weather phenomena were considered closet infidels, who were debasing the mysterious ways of God by so measuring their effects. It was as if someone were using a magnifying glass to see the beauty in a Filippo Lippi painting of the Madonna.

This popular disdain could be intimidating. One of Parker Cleaveland's fellow weather watchers in Maine felt the sting of public opprobrium when it was revealed that he had been keeping weather-related records. In February 1817, this man — Moody Noyes — wrote to Cleaveland that he had detected a "disposition from a certain quarter, to treat all the observations made here with ridicule or something worse." His name having twice been mentioned in a manner that had "displeased and disgusted" him, Noyes had decided to "give up stargazing for the present" and pass on his thermometer to a less timorous gentleman.[3] For all of these various reasons, predicting the weather in the early nineteenth century remained largely a matter of guesswork.

The proliferation of newspapers changed this situation. The ascendant republican ethos put a premium on providing the American people with timely and relevant information. To be well informed was a civic basic need, but also a means of empowering the citizenry, ending the monopoly of a privileged few on knowledge about political, economic — and scientific — developments. Fueled by growing prosperity, the public's appetite for news led to the establishing of hundreds of daily and weekly papers.[4] From a mere ninety in 1790, their number soared to 370 two decades later. Most of these new newspapers were aligned with the Republicans.[5] As part of their

---

[1] Quoted in Robert H. Eather, *Majestic Lights, The Aurora in Science, History, and the Arts* (Washington, DC: American Geophysical Union, 1980), 93.

[2] *Agricultural and Economical Almanack*, [not paginated].

[3] Letter of Moody Noyes to Parker Cleaveland, 25 February 1817, M34, Folder 27, Parker Cleaveland Collection, George J Mitchell Department of Special Collections and Archives, Bowdoin College, Brunswick ME.

[4] The overwhelming majority of newspapers were published on a weekly basis.

[5] Those with that editorial slant doubled from eighty to 160 during the same period, so that by 1810 more newspapers were advancing Republican positions and achievements than Federalist ones. Jamie L. Carson and M.V. Hood III, "The Effect of the Partisan Press on U.S. House Elections, 1800-1820," (Paper presented at the annual meeting of the Midwest Political Science Association, Chicago, 2008), Fig. 4, "U.S. Newspapers by Type,

educational mission, they broadened their coverage to places and events far from where their readers lived. They printed summaries of Congressional debates, letters and reports from the Western territories, and articles about the Napoleonic Wars and other military and political events in Europe. Even though some international stories did not show up in American papers until several months after the events described had taken place, they still made it possible to keep up with what was going on around the world. In rural East Windsor, Connecticut, for example, Thomas Robbins could closely follow the progress of the conflict between England and France.[1] This expanded news coverage gave American readers a sense of belonging to a larger world.

Before 1816, few newspapers printed weather-related information. Occasionally, a publication in Boston or New Haven might include some local meteorological readings, provided by natural philosophers like Yale's Jeremiah Day. But day- or week-old readings were of little practical value and, hence, only published when the weather was unusually hot or cold. Reports from afar were even less newsworthy as they were not considered relevant to conditions closer to home. However, as with political reporting, interest in the bigger picture gradually increased. By 1800 a few papers cited temperatures taken at more than one location, making note of their similarities. For example, the Boston-based *Columbian Centinel* printed readings from Hartford and Salem on the same August day.[2] In 1806, the *Connecticut Gazette* ran a column listing the average temperatures, as well as highs and lows, in New London during the month of January for the previous four years. Accounts of severe weather in Europe and in other parts of the United States stirred speculation about how — if at all — these abnormalities might be related. In October 1815, the editor of *The Gazette of the United States* wondered why, at the same time a hurricane had been raging in Boston, the weather where he was, in Philadelphia, had remained "clear and mild." "It would be accomplishing a desirable object," he wrote, "if men of leisure and observation in different and remote parts of the United States, would habitually make accurate meteorological observations, and at short intervals publish them. From a comparison of such documents, some scientific deductions might, [*sic*] perhaps result."[3]

The odds of this happening improved greatly after the cessation of hostilities with Britain. At that time almost all scientific instruments were

---

1800-1820," 11. http://www.polls.uga.edu/APD/Carson%26Hood.pdf.
[1] For example, Robbins noted in his diary in the spring of 1796: "By the latest accounts, very little prospect of peace in Europe very soon. The democratical [*sic*] interest very lawless." Robbins, *Diary*, entry for 8 June 1796, 11.
[2] Untitled article, *Columbian Centinel*, 13 August 1800.
[3] "The Weather in Philadelphia," *Merrimack Intelligencer*, 7 October 1815.

manufactured abroad, primarily in England.[1] Once the fighting was over, shipments of thermometers resumed, as can be seen in the newspaper advertisements selling these products after 1814.[2] This availability of "Fahrenheit's thermometer" enabled more well-to-do amateur weather watchers to keep their own records. In the years leading up to 1816, they did so out of sheer curiosity. However, when temperatures began fluctuating wildly in the spring of 1816, this interest took on more urgency. Measurements made across the United States and then printed in newspapers gave the reading public an overview of this bizarre weather.[3] The *Providence Patriot*, in Rhode Island, was the one of the first publications to do this, reprinting articles on conditions in states from Vermont to Pennsylvania in its June 22nd issue.[4] Close observers of the weather in New England began to realize that the calamitous cold they were experiencing was part and parcel of a much larger phenomenon.

To understand it, more meteorological data had to be collected. This was what a letter writer in Kennebec, Maine, proposed that July, in response to the "very gloomy" situation facing farmers in the District. With severe drought and extreme cold having decimated their potato, wheat, corn and other crops, and starvation looming, he expressed his own despair ("What is to come of us?") and implored editors in other states to end this "dearth of political intelligence" on the unfolding situation by sharing information.[5] His impassioned request achieved its desired effect. This letter was widely reprinted, and the publication of weather reports from various locales increased greatly over the rest of the summer. More facts promoted conjecture about what was responsible.[6] In late August, the nationally distributed

---

[1] London's Pastorella and Co. was supplying thermometers in New York at this time. See advertisement for their product, *National Advocate* (New York), 30 July 1817. The Boston-based New England Glass Company — which became the largest manufacturer in the world in this field — did not start to produce thermometers until 1818.

[2] The price for these instruments remained sufficiently high that only a few could afford them. Contemporary reports in New England newspapers of stolen thermometers, offering rewards, indicate how highly valued they were.

[3] See, for example, "The Weather," *Providence Patriot*, 11 March 1817. In this article, the Rhode Island newspaper announced it was henceforth going to print monthly tabulations of meteorological observations, including temperature, wind direction, and quality of weather, "which we hope all accurate and intelligent observe will follow . . ."

[4] "Snow Storm," *Providence Patriot*, 22 June 1816.

[5] "Domestic: A Scarcity of Crops," *Lancaster Journal* (Lancaster PA), 24 July 1816. Reprinted from the *Boston Gazette*.

[6] Theories about this apparently large-scale climate shift had surfaced earlier in the summer. For example, a July article in a Vermont newspaper, commenting on weather reports received from all over the country, had described the summer as without parallel in living memory and then discussed possible explanations, including the unusually large amount of ice

Niles' Weekly Register summarized conditions from Canada to Virginia and then reminded readers that a cooling trend had been noticed for several years. The author ventured that recent earthquakes might be to blame.[1] That same month, a writer for the *Richmond Commercial Compiler* pointed out that the clearing of forests in Europe had been followed by warmer temperatures there and opined that the same activity would eventually produce the same change in North America as well. He urged his readers to adopt this long-range outlook: "We must have an eye to a long succession of seasons, and take the average of all — this is the only means by which we come at the truth."[2] Other commentators noted that Herschel's theory that a high frequency of sunspots resulted in warmer terrestrial temperatures now appeared discredited, since the spots observed in 1816 were unusually large and numerous.[3]

Most helpful — and also confounding — to those trying to figure out why the weather had turned so unseasonably cold were reports from Europe. These had started arriving before New England had suffered its late spring frosts. A man in Boston received a letter from Bordeaux describing an erratic pattern in the weather there: the temperature had dropped thirty degrees overnight, so that "sourtouts and winter garments were as common . . . as in December."[4] By the middle of July, accounts of heavy winter snowfalls across the Continent were published in numerous New England papers: hills in Scotland had been buried under the deepest snows in two decades.[5] More baffling, the snow continued to fall, in some places, when the start of spring was only a few weeks away.[6] Toward the end of summer, Americans could read about widespread food shortages, rioting, and fears of starvation in England, Germany, Sweden, and other countries — all caused by relentless downpours which had ruined crops. Europe, in short, was dealing with a climate crisis very much like what was happening on the other side of the Atlantic. This awareness of unusual weather thousands of miles away was virtually unprecedented. (The only parallel situation was in 1740–41, when

---

seen floating in the Atlantic, sunspots, and the clearing of forests. None of these held up to closer scrutiny, the writer concluded. He could only end with a resigned shrug: "Neither our leisure nor our limits will admit of a farther discussion of this subject at this time." "The Season," 17 July 1816, *Reporter* (Brattleboro VT).

[1] "Climate of the United States," *New-Jersey-Journal* (Elizabethtown NJ), 20 August 1816. Reprinted from *Niles' Weekly Register*, 10 August 1816.

[2] "On the Climate," *National Daily Intelligencer*, 27 August 1816. Reprinted from the *Richmond Commercial Compiler*.

[3] "The Climate," *Connecticut Journal*, 17 September 1816. Reprinted from the *Richmond Compiler*.

[4] Untitled article, *Connecticut Courant*, 11 June 1816. This letter was written on April 5, 1816.

[5] "European Snow," *Vermont Gazette* (Bennington), 16 July 1816.

[6] "Munich, June 20," *Daily National Intelligencer*, 24 August 1816.

England and North America had both endured a bitterly cold winter, but then news traveled more slowly.) Learning of this common catastrophe affected Americans and Europeans profoundly, in two ways.

First of all, it was now undeniable that the forces producing climatic changes were immense — large enough to destroy crops, livestock, and human lives over a wide swath of the planet. Their scale indicated that the weather was caused by equally large, possibly even extraterrestrial events — such as sunspots. But this new perspective raised several unsettling questions. If the earth's climate was subject to the vagaries of celestial bodies, could Newton's notion of a well-ordered, mechanistic universe still be valid? If such remote phenomena, occurring so randomly, determined whether crops would be abundant or fail, how could humanity feel secure? The notion of beneficial bonds between man and Nature might be a fiction. Such a conclusion conjured up images of a helpless human species alone in a vast, indifferent void.

Furthermore, if natural forces like storms and droughts occurred without any higher purpose, what did this say about God's relationship with the beings he had made in His own image? Previously, it had been possible to interpret natural disasters as signs of divine wrath. But this logic only made sense when there was some connection between those who suffered and their moral or religious failings. For instance, when a lightning bolt knocked Martin Luther to the ground in 1505, he had good reason to interpret this as God's command that he should abandon his plans to enter the law and instead become a monk. But once it became apparent that these acts of "punishment" befell sinners and the devout indiscriminately, this conviction was shaken. Doubts about the Almighty's intent had already been raised — for example, in 1735, when a powerful bolt had struck a Congregationalist church in New London, Connecticut, setting the roof ablaze and causing burning beams to rain down on members of the terrified congregation. A particularly pious young man had perished. Afterwards the minister, Eliphalet Adams, had delivered a sermon contending that this "awful and terrible thing" had been done out of God's righteousness, to revive a feeling of awe at His presence.[1] But explanations of this sort did not always sit well with the afflicted: a natural cause seemed more plausible. The Lisbon tsunami and earthquake had lent credence to this point of view. New England's experiencing a quake a few weeks later had added more weight to the belief that natural forces were causing these upheavals. Coming on the heels of these traumatic events, the "year without a summer" prompted more efforts to figure out what they were.

---

[1] Eliphalet Adams, *God Sometimes Answers His People, by Terrible Things in Righteousness* (London: T. Green, 1735), 8-10.

A major step in that direction was taken by the Philadelphia Society for Promoting Agriculture, when it voted in October 1816 to begin compiling weather-related statistics from all over the United States in order to predict future famines.[1] But the most important weather data-collecting effort was undertaken by a 59-year-old ex-city clerk, ex-mathematics tutor, ex-printer, ex-lawyer, and ex-university president named Josiah Meigs. A native of Connecticut, Meigs had graduated from Yale in 1778 — at the height of the Revolution — with a class whose members included Daniel Webster and Oliver Wolcott, Jr. Meigs had stayed on for several years to teach mathematics, natural philosophy, and astronomy while he studied law. He also married Clara Benjamin, the daughter of a colonel in the Revolutionary Army, by whom he would have nine children. In 1783 Meigs was chosen to be New Haven's first city clerk. He also opened a printing shop in that city, publishing the *New Haven Gazette*. Meigs used this editorial platform to drum up local support for adopting the Constitution and, subsequently, for advancing Jeffersonian principles of government. (Meigs was almost viscerally hostile to any form of elitism and upbraided the Adams administration for being "tyrannical" and an "infamous junto."[2]) In 1788 he sailed to Bermuda, to oversee the legal affairs of several of his Connecticut clients. But Meigs' combativeness and loose talk soon got him into hot water with American authorities, and he was charged with treason. (He was ultimately acquitted.) Seeking a more quiet life, he returned to New Haven and assumed a chair of mathematics and natural philosophy at his alma mater. This academic sojourn ended shortly after Jefferson's election to the presidency, when Meigs was named president of a new college, Franklin, in Georgia (later the University of Georgia). A few years after arriving in Athens, he began gathering information on the state's climate, to demonstrate its healthiness and thereby attract more residents.[3]

Meigs envisioned this Georgian college developing into another Yale, and so he implemented a classical curriculum, supplementing it with courses in mathematics and the natural sciences. A man without any sectarian affiliation, he soon ran afoul of the college's trustees, who wanted Franklin to become a Christian institution.[4] This was somewhat ironic, given that the college had been named for a celebrated Deist. Meigs and one of the trustees locked

---

[1] John D. Post, *The Last Great Subsistence Crisis in the Western World* (Baltimore: Johns Hopkins University Press, 1977), 13.
[2] Quoted in William Montgomery Meigs, *Life of Josiah Meigs* (Philadelphia: J.P. Murphy, 1887), 66.
[3] "This Day in Athens: Wednesday, April 28, 2010." http://accheritage. blogspot.com/2010/04/28-april-1806-josiah-meigs-records-last.html.
[4] Meigs branded the board "a damned pack or a band of Tories and speculators." Quoted in F.N. Boney, *A Pictorial History of the University of Georgia* (Athens: University of Georgia Press, 1984), 11.

horns over the college's educational philosophy, and, in 1810 the board forced Meigs to resign. He stayed on for another year as a professor of chemistry, but was then fired from that position, for alleged "great misconduct" — namely, public criticism of the board.[1] Meigs brought his family north the following year, to the nation's capital, where he taught "experimental philosophy" at the Columbian Institute (later George Washington University). President Madison then offered him the post of surveyor general in the General Land Office, to oversee the distribution of public acreage in newly acquired Western territories. Meigs moved to Cincinnati to take up these duties, but hostilities with British forces and their Native American allies prevented him from carrying them out.[2] In the fall of 1814 he arranged to swap jobs with his friend Edward Tiffin, a physician and former governor of Ohio, and become commissioner of the General Land Office, in Washington.[3] Back east Meigs kept up his keen interest in meteorology. In 1816, he observed an unusually large sunspot through his telescope and speculated that this was some kind of "Solar Volcano." [4] Since his days in Bermuda, Meigs had been fascinated by the weather and had kept detailed temperature records.[5] The extremely cold during "the year without a summer" intrigued him. A Cincinnati acquaintance of his, Dr. Daniel Drake, had been collecting information on the weather since 1806, and his meticulously kept ledgers and recent book on the relationship between climate and disease induced Meigs to take on a much grander project.[6] He was also motivated by the wartime decision of the Army Medical Department to keep "weather diaries" and his own previous gathering of meteorological data, in Bermuda and in

---

[1] Accused of holding the board in contempt, Meigs dismissed these charges as "infamous and cold-blooded calumny." In a letter to the board, he wrote: "I have always believed that knowledge pure and unveiled (either politically or religiously) was the true basis of happiness of individuals or of nations. Idiots and fools alone will consent to unreasonable laws, and dishonest men alone will declare their belief of that which their judgment and conscience declare to be false." He contended he had become a target "for the shafts of those who dread the true light of science." See letter of Meigs to Board of Trustees, Franklin College, 9 August 1811. http://www.libs.uga.edu/hargrett/archives/meigs/response.html.

[2] Malcolm J. Rohrbaugh, *The Land Office Business: The Settlement and Administration of American Public Lands, 1789-1837* (New York: Oxford University Press, 1968), 57.

[3] Ibid. 57.

[4] Letters of Josiah Meigs to Dr. Daniel Drake, 30 April and 30 September 1816, Meigs, *Life*, 96, 98.

[5] Ibid., 81.

[6] In 1810 Drake had published *Topography, Climate, and Diseases of Cincinnati.* Five years later he wrote a book describing the environment and unusual phenomena, such as earthquakes, experienced in Cincinnati and surrounding Miami County. See Thomas W. Schmidlin and Jeanne A. Schmidlin, *Thunder in the Hole: A Chronicle of Outstanding Weather Events in Ohio* (Kent OH: Kent University Press, 1996), 81.

Georgia.[1] In 1807, the first Surveyor General, Colonel Jared Mansfield, had begun making weather observations, but on a limited scale. [2] Now, in the spirit of Thomas Jefferson, Meigs planned to make use of his public office to increased scientific understanding of the weather. He hoped that climate-related facts would spur Western settlement (and the sale of government lands) by identifying places where healthy and favorable growing conditions existed.[3] In describing his project to Land Office officials, Meigs expressed hope that it would "increase our physical knowledge of our country."[4]

In February, 1817 he wrote to Drake, outlining this plan. He intended to instruct the heads of regional land offices in cities like Detroit, St. Louis, and New Orleans to keep logs of temperature, barometric pressure, rainfall, and wind direction and speed, and to send monthly reports to Washington. Easily collected, at little expense, these data would constitute a longitudinal record of the country's weather and reveal patterns in its changes.[5] This would be the first coordinated attempt to collate and analyze meteorological facts on such a large scale. However, despite his lobbying of several influential members of Congress, Meigs failed to secure funding for this undertaking. He wrote to Drake, expressing his frustration that the American public remained as "ignorant on this subject [the weather] as the Kickapoos [a Native American people] now are . . ."[6] Undaunted, Meigs decided to proceed without government backing: he mailed blank forms to his regional land offices in April, 1817, asking them to record weather information three times a day. He also requested his deputies to make note of such seasonal events as the unfurling of plant leaves, budding of flowers, migration of birds and fishes, hibernation of animals, falling of objects from the sky, and changes in human health, including "epidemic and epizootic distempers."[7] But without funds to purchase the required instruments, the regional offices could not fully comply.[8] Nonetheless, some meteorological data were regularly sent to the General Land Office over the next two decades. As a result of this input,

---

[1] The Army continued to collect meteorological information until 1890, when it turned over this responsibility to the Weather Bureau. Robert D. Ward, "A Short Bibliography of United States Climatology," *Transactions of the American Clinical and Climatological Association* 34 (1918): 2.

[2] Campbell Bayard, "The Government Meteorological Organisations in Various Parts of the World," *Quarterly Journal of the Royal Meteorological Society* 25:110 (April 1899): 123.

[3] In pursuing these goals, he was following Jefferson's lead.

[4] Josiah Meigs, "Circular to the Registers of the Land Offices of the United States; General Land-Office, April 29, 1817," *Columbian Register*, 10 May 1817.

[5] Letter of Josiah Meigs to Dr. Daniel Drake, 1 February 1817, Meigs, *Life*, 81-82.

[6] Letter of Meigs to Drake, 13 June 1817, Meigs, *Life*, 82.

[7] Meigs, "Circular to the Registers."

[8] Eric R. Miller, "The Evolution of Meteorological Institutions in the United States," *Monthly Weather Review* 59:1 (January 1931): 1.

Meigs was able, in December 1818, to map a cold front extending all the way from Georgia to Maine.[1] This was the first time that a weather system of this magnitude had been so clearly delineated. Each month for more than three years, Meigs plotted statistics from across the country on a map mounted his office. This visual representation enabled him, again for the first time in history, to see how weather disturbances moved through time and space. In 1821, the commissioner submitted a paper outlining his "geometric exemplification" of Washington's weather to the American Philosophical Society, but it was rejected for publication.[2] Meigs died the following year, his pioneering work largely unrecognized.

But it would have a major impact on the development of meteorology. Along with the German mathematician Heinrich Wilhelm Brandes (who, in 1817, also drew up weather maps and invented isobars — lines connecting locations with equal barometric pressure[3]), Meigs paved the way for longitudinal studies of weather changes. In 1824, Simeon DeWitt, the New York Surveyor, set up weather stations at several branches of the state university. They generated records until the middle of the Civil War. In 1834, James Pollard Espy oversaw a joint effort by the American Philosophical Society and the Franklin Institute to track the path of storms in Pennsylvania, furnishing every county with instruments to gauge the strength and direction of winds, temperature, precipitation, and atmospheric pressure. At Espy's urging, the Pennsylvania Lyceum petitioned Congress for funds to establish the nation's first weather service. But these were not obtained until 1849, from the Smithsonian Institution. In the ensuing decades, in both the United States and Europe, steady progress was made in mapping the weather and studying its causes.[4] By early in the twentieth century it finally became possible to make reasonably reliable forecasts, drawing upon new knowledge of the physics of the atmosphere.

It would be an exaggeration to claim that all these advances stemmed from the reports which reached Josiah Meigs's desk in Washington beginning in 1817. As if often the case in scientific research, people at different places around the world used the same instruments and strategies at approximately the same time to discover how weather patterns develop and spread. Meigs happened to be in the forefront of this inquiry at a propitious time — when

---

[1] Meigs, *Life*, 84.
[2] Ibid., 113.
[3] These were first published in 1820 in his *Beiträge zur Witterungskunde* (*Contributions to Meteorology*).
[4] In most European countries, as well as in many South American and Asian countries, meteorological services came into existence in the 1840s and 1850s. Switzerland was the earliest country in Europe to establish one, in 1824. Bayard, "Meteorological Organisations," 101. A U.S. government weather bureau was not created until 1870.

an expanding United States, a growing penchant for scientific inquiry, and relatively sophisticated instrumentation made it possible to achieve major breakthroughs. It would be equally misleading to assert that Meigs undertook his seminal record-collecting solely in response to the calamitous cold spells of the previous year. He had been fascinated by the weather for many years before then. Still, what took place during the "year without a summer" no doubt made the need to chart weather systems and predict their movement much more pressing. It is hard to imagine that Meigs was not greatly puzzled by the bizarrely oscillating temperatures experienced during the spring and summer of 1816. It is very likely — if not provable — that his desire to collect weather data from all over the country was a direct response to this disconcerting change in climate.

The larger reality was that — around the time that New England and northern Europe were enduring these unprecedented cold spells — major Western thinkers were questioning the long-standing belief that the weather was controlled by God. Just as the great Lisbon earthquake of 1755 had shaken confidence in the Deity's omnipresence, so did this year of failed crops and famine strongly suggest that at least some earthly events arose from purely physical causes. Natural philosophers, statesmen, and men of letters henceforth became more interested in investigating these. More importantly, the weather of 1816 gave many ordinary folk reason to wonder if the universe was, indeed, governed by a divine being. The possibility that this might not be the case stirred feelings of desolation and abandonment. For others, this thought was liberating. If God did not control all things, then perhaps human beings possessed the freedom to shape their own destiny. Perhaps it was they, and not the Almighty, who could hurl the thunderbolts.

Chapter 5. From Ploughshares to Power Looms

Those who labor in the earth are the chosen people of God.
— Thomas Jefferson, 1785

He, therefore, who is now against domestic manufacture, must be for reducing us ... to be clothed in skins, and to live like wild beasts in dens and caverns.
— Thomas Jefferson, 1816

Poor harvests in 1816 focused attention on New England's underlying economic weakness — dependence on unreliable and increasingly less productive agriculture. To remain competitive with other parts of the nation and keep young families from migrating to the West, new kinds of opportunities had to be developed. This realization that farming could not be counted on spurred the region's transformation into a major manufacturing center.

As the wagons lurched off toward the Western mountains — dogs yapping, whips cracking, babies bawling, pans rattling, lambs bleating, women weeping, men cajoling, girls giggling, boys whistling — the people staying behind could only passively stare and wave back. For better or worse, the land was now theirs. It was up to them to keep their way of life going. Tilling these hard, unyielding soils had never been easy, and now it seemed harder than ever. The cruel frosts of 1816 had damaged not only their Indian corn, wheat, rye, vegetables, and fruit trees, but also their confidence and innate stoicism. God's failure to make the earth bountiful had tested their faith. Still, they had survived the coldest spring and summer ever known by Europeans on this continent and so now, as the first intrepid shoots nosed up through the moist earth, these descendants of John Winthrop had grounds to feel relieved. With the early morning sun warming their upturned faces, they once again headed out to their gardens and fields with easy, loping strides, with quiet hopes that crops would grow this year in abundance.

But worries persisted. Was it possible — as some were murmuring — that chilly, barren summers lay ahead? Would New England remain habitable? Or was God sending a signal that this land of their ancestors was no longer where they should build the "City upon a Hill," as Winthrop had admonished them to do? For ice and frost were not their only enemies. People were leaving for a Western Promised Land because this one was used up. After planting the same fields for centuries, New England farmers had sapped them of their nutrients. A shortage of arable acreage discouraged experimentation with new crops.[1] By 1800, the typical New England farm had only three acres under cultivation, with the remaining forty-odd acres set aside for grazing animals or for firewood.[2] In the past, cash crops had complemented this subsistence farming. As far back as the seventeenth century, wheat had been grown in the Connecticut Valley, ground into flour in local mills and then sold locally or hauled off to market in cities like Boston.[3] But a deadly fungus and then, in the 1800s, an insect infestation (Hessian flies) had put an end to virtually all commercial wheat production.[4] Thereafter, barrels of flour were shipped up the Connecticut River to meet the local demand for bread.[5]

Generally speaking, farmers produced only what they and their families needed. In addition to growing their food, they wove wool to make rough shirts, blouses, pants, and skirts; sawed, chiseled, and sanded boards to make tables, cabinets, and chairs; tanned cowhides to turn them into shoes, belts, coats, and aprons; and hardened soft clay into bowls, cups, and dishes. There was virtually no cash economy and only a limited barter system. If one farmer had some surplus corn or apples, he might exchange these for, say, nails or

[1] Christopher Clark, *The Roots of Rural Capitalism: Western Massachusetts, 1780-1860* (Ithaca, London: Cornell University Press, 1990), 61. Clark notes that, by 1800, many New Englanders had become landless as acreage had grown scarce — and too expensive for them to buy.

[2] John M. Murrin, et al., *Liberty, Equality, Power: A History of the American People*, vol. 1. 4th ed. (Belmont CA: Thompson/Wadsworth, 2005), 267.

[3] Judd, *History of Hadley*, 354. Grist mills had been operating in New England since the early 1600s.

[4] Jamie H. Eves, "'And Then What Shall We Do?' How Black Stem Rust, Hessian Flies, and Wheat Midges Destroyed Wheat Farming in Connecticut." http://www.millmuseum.org/Mill_Museum/Decline.html. Cf. Timothy Dwight, *Travels in New-England and New-York*, vol. 1 (New Haven: Timothy Dwight, 1821), 49.

[5] In the middle of the eighteenth century, Boston was also receiving shiploads of corn and flour from the Mid-Atlantic states. Peter Kalm, *Travels into North America*, vol. 1, trans. John R. Forster (London: C. Lowndes, 1772), 253. Starting in the 1820s, tobacco leaves were grown in the Connecticut River Valley for domestic use and export as cigar wrappers.

wrought-iron tools from a neighbor who happened to be a blacksmith. But, in the main, each adult male was expected to take care of himself and his family. Thus, the Congregationalist minister Thomas Robbins spent a great deal of time cultivating his own vegetables and fruits in his East Windsor, Connecticut, garden.[1] Although he was a Yale graduate, a doctor of divinity, and a widely-read man of letters, he could not expect his parishioners to leave freshly-picked peas on his doorstep or pay him a salary sufficient to live on. Survival was up to him.

This meant there was little division of labor, other than between the sexes. Few could make a living from only doing one thing well.[2] In small towns and villages, carpenters, coopers, blacksmiths, and potters constituted the rare exceptions to this general rule. All the rest had to be jacks of all trades. Ninety percent of New England's population lived on small farms, leading lives restricted by a short growing season, depleted soil, and inadequate tools. The routines of tilling, seeding, and harvesting had not changed for generations. Indeed, if a farmer from the mid-eighteenth century could have been teleported onto the same plot of land seventy five years hence, he would have found himself very much at home.

This continuity was also sustained by the farmers' suspicion of "newfangled" ways. Upholding tradition prevented the region's agriculture from becoming more efficient or productive. It lagged far behind England's in introducing new methods and tools, such as the iron plow.[3] While some Yankees were demonstrating their legendary ingenuity by inventing new ways of making clocks, dyeing hats, spinning wool cloth, and picking cotton, almost no important innovations originated on the farmlands of Connecticut, Massachusetts, Vermont, New Hampshire, or Rhode Island.[4]

---

[1] His fellow clergyman, Yale president Timothy Dwight, annually planted corn in his New Haven garden a dozen different times during the growing season. Dwight, *Travels in New-England and New-York*, vol. 1, 49.

[2] By 1810, very few New Englanders were engaged fulltime in manufacturing. Most manufactured products, such as cotton and wool cloth, were made in individual households. Percy W. Bidwell, "Rural Economy in New England at the Beginning of the Nineteenth Century," *Transactions of the Connecticut Academy of Arts and Sciences* 20 (April 1916): 275.

[3] So admitted Yale president Dwight in his *Travels in New-York and New-England*, vol. 1, 108. The iron plow was developed in England in the 1730s, but was not introduced in the United States until 1797, when a patent was issued to Charles Newbold, a New Jersey blacksmith. Even then, many American farmers put off switching to this more efficient plow because of fears that iron might poison their soils. Walter Ebeling, *The Fruited Plain: The Story of American Agriculture* (Berkeley: University of California Press, 1979), 91. Only in the 1840s was it widely adopted. Peter D. McClelland, *Sowing Modernity: America's First Agricultural Revolution* (Ithaca, London: Cornell University Press, 1997), 43.

[4] 9 Eli Whitney, a farmer's boy from central Massachusetts, had gone south to make his mark and settled on a plantation in Georgia where he invented the cotton gin.

New England farmers made no effort to breed sheep which produced the higher quality of wool required to make clothing, even though demand for these manufactured products was growing.[1] Instead, they continued to produce only rough, homespun yarn.[2] Furthermore, they saw no point in improving how they plowed their fields, tended their crops, or raised their animals, for they could not expect to make a profit by doing so. There was only a small urban market for their goods.[3] Few people lived in the region's slowly growing cities.[4] In 1810, only three New England towns — Boston, Providence, and Salem — had as many as ten thousand residents, and their combined population of fifty-six thousand accounted for just 6.9 percent of the total.[5] And, even in the cities, many people continued to grow their own food.[6]

With no buyers for their foodstuffs, New England farmers were caught in a bind.[7] It made no economic sense for them to produce more, but if they didn't, they would gradually be forced off their land. Soaring land prices, smaller available plots, fluctuating and modest harvests, and now uncertain

---

[1] Bidwell, "Rural Economy," 340.

[2] During the colonial period, wearing homemade clothing was also an act of defiance against the British, who were trying to force New Englanders to buy cloth imported from England. In the 1760s, Harvard seniors donned homespun wool for their graduation ceremonies to express their opposition to this policy. Lawrence A. Peskin, *Manufacturing Revolution: The Intellectual Origins of Early American Industry* (Baltimore: Johns Hopkins University Press, 2003), 43.

[3] Dwight, *Travels*, 268.

[4] Between 1810 and 1820, Salem grew by only some 120 persons, Gloucester by fewer than five hundred, and Portsmouth, by almost four hundred. New London virtually did not expand at all, while New Bedford lost thirty percent of its population, down to 3,947 from 5,651. In the previous decade, Hartford grew by less than nine percent, but other cities — Providence, Boston, New Haven, and New Bedford — increased in size significantly. See Campbell Gibson, "Population of the One Hundred Largest Cities and Other Urban Places in the United States, 1790 to 1990." Population Division, U.S. Bureau of the Census, Working Paper No. 27, June 1998. http://www.census.gov/population/www/documentation/twps0027/twps0027.html.

[5] Percy W. Bidwell, "Population Growth in Southern New England, 1810-1860" *Publications of the American Statistical Association* 15:5 (December 1917): 815. However, by 1860, 36.5 percent of the region's people were living in towns and cities with more than ten thousand residents.

[6] Dwight estimated that five out of eight persons living in New Haven (pop. 10, 071) in 1810 were engaged in producing cotton goods. But over half the population still took part in farming. Bidwell, "Rural Economy in New England," 281, 292.

[7] For an opposing view of the growth of New England markets, see Winifred Rothenberg, *From Market-Places to a Market Economy: The Transformation of Rural Massachusetts, 1750-1850* (Chicago and London: University of Chicago Press, 1992). Drawing upon her close reading of extant farmers' journals, Rothenberg concluded that the building of better roads made it easier for Massachusetts farmers to take their produce to markets in the state and beyond starting in 1800.

growing seasons were making agriculture unviable. Hence, the exodus westward, which had been going on for decades, kept gaining momentum. Between 1790 and 1820, an estimated eight hundred thousand people left New England for other parts of the country — mostly to the former Northwest Territory.[1] This number was equal to half the number of persons who would otherwise have populated the region.[2] The great majority came from rural, inland areas — an indication that farm difficulties were the main "push."[3] In addition, while the populations of Massachusetts and Connecticut grew by roughly sixteen percent during each of the first four decades of the century, the number of persons engaged in farming remained unchanged.[4] Farmers were becoming a smaller part of the labor force, mainly because so many were leaving these two states. To make matters worse, the largest group emigrating was young families, robbing the region of those could assure its agricultural future.

Struggling, indebted farmers had only two choices: make a profit off their own lands by specializing in one crop or form of animal husbandry — or go somewhere else. The latter had long been — as John Tudor put it in his *Letters on the Eastern States* (1820) — the "easy remedy."[5] But recent developments across the Atlantic suggested another option. Since the mid-eighteenth century, water-powered cotton mills in the north of England had been turning out large quantities of cloth. Thanks to the introduction of the spinning jenny and other machinery around the time of the Revolution, they had upped production — and lowered costs — considerably. Imports of raw cotton from the Southern states allowed Britain to expand its textile industry rapidly. The value of its cotton goods soared from some $36,000,000

---

[1] Connecticut alone lost some 238,000 persons over these thirty years. David R. Meyer, *The Roots of American Industrialization* (Baltimore: Johns Hopkins University Press, 2003), 41. Between 1810 and 1860, farming communities in Massachusetts, Connecticut, and Rhode Island would see their populations decline by nearly five percent.

[2] This would have happened as a result of natural increase and immigration. Bidwell, "Rural Economy," 387.

[3] Census figures for agricultural counties show steeper population declines during this thirty-year period than do coastal cities and manufacturing centers like New Haven, Providence, and Boston. For example, three inland, rural counties in Connecticut (Litchfield, Windham, and Tolland) had a net gain of only 0.6 percent from 1790 to 1810, whereas New Haven, Fairfield, and Hartford counties increased their total population by 16.8 percent. Bidwell, "Rural Economy," 387. David Meyer notes that many of these farm migrants came from areas where the land quality was poor. He thus argues that New England did not experience a "widespread rural decline" early in the nineteenth century. Meyer, *Roots of American Industrialization*, 34.

[4] In 1800, Connecticut had 50,400 farm workers, and Massachusetts 73,200. By 1830, the figures were 55,910 and 78,500, respectively. Over that thirty-year period, the combined population for the two states rose from 673,847 to 1,047,214 — or sixty-four percent.

[5] Tudor, *Letters on the Eastern States*, 202.

(in today's dollars) in 1770 to $630,000,000 three decades later.[1] By then, more than half of the cloth produced in England was being exported to the United States.[2] Seeing their former colonial master grow rich from textile manufacturing, some American businessmen wondered why their own country could not prosper from the same enterprise. Prominent public figures took up this cause — most famously, Alexander Hamilton, who, while serving as Secretary of the Treasury, presented a report to Congress in December 1791 arguing that manufacturing could be just as valuable to the country's economy as agriculture.[3]

American manufacturing had existed since the colonial era. Glass, iron tools, porcelain, furniture, silverware, and soap, in addition to woolen cloth, had been produced, mostly in private homes and in limited quantities. Making such household items had taken on patriotic overtones after the British began forcing the colonists to pay duties on imports. Notables like George Washington and Benjamin Franklin had advocated the domestic production of cloth and other goods to end dependence on England. Many colonists urged their friends and neighbors to shear sheep instead of eating them and spin the fleece into wool. Nationalistic pride led Washington to don a broadcloth coat and silk stockings made domestically at his first inaugural.[4] After the Revolution, factories were built for making candles, nails, and cotton goods.[5] Both Franklin and Hamilton were instrumental in establishing societies to promote manufactures.[6] They foresaw that America's future greatness would require the growth of a vibrant manufacturing sector.

---

[1] Cotton shipped to England increased from seven percent of total U.S. exports in 1800 to forty-one percent by 1830. Jean M. West, "King Cotton: The Fiber of Slavery," *Slavery in America*.

[2] Brian Arthur, *How Britain Won the War of 1812: The Royal Navy's Blockades of The United States, 1812-1815* (Woodbridge, Suffolk: Boydell Press, 2011), 47. In 1805, cotton accounted for forty two percent of British exports.

[3] Alexander Hamilton, "Report on Manufactures," 5 December 1791, *Selected Writings and Speeches of Alexander Hamilton*, ed. Morton J. Frisch (Washington DC: American Heritage Institute for Public Policy Research, 1985), 278-319. Hamilton made note of England's "immense progress" in cotton manufacturing. He also pointed out that agricultural exports were not a reliable source of income. Hamilton subsequently founded a "Society for the Establishment of Useful Manufactures," with the initial goal of operating a cotton mill on the Falls of the Passaic in New Jersey. J. Leander Bishop, *A History of American Manufactures from 1608 to 1860*, vol. 2 (Philadelphia: Edward Young, 1864), 31.

[4] Shortly before taking the oath of office, Washington had visited the highly regarded Hartford Woolen Company and had this coat made from material he had admired there. Zakim, "Sartorial Ideologies," 1558.

[5] Peskin, *Manufacturing Revolution*, 61.

[6] Under Franklin's coaxing hand, the Pennsylvania Society for the Encouragement of Manufacturers and the Useful Arts came into being in 1787. Four years later, in New York, Hamilton helped create a similar body, the Society for Establishing Useful Manufactures.

But this vision was not widely endorsed. Many Americans rejected a shift to manufacturing as a betrayal of the country's agrarian ways. They feared that urban life might sever a vital emotional connection to the land and the rugged individualism fostered by it. Opposition to Hamilton's 1791 report was particularly strong among Democratic-Republicans. They felt that concentration of wealth and power in the hands of factory owners and banks would create income and class disparities in the new nation and reduce yeoman farmers to wage workers, stripped of the human dignity which came from tilling the soil. Jefferson believed industrialization would spell the end of democracy. In his 1787 *Notes on the State of Virginia* he lauded the independent American farmer as the guardian of the nation's values and traditions: "Those who labour in the earth are the chosen people of God, if ever he had a chosen people, whose breasts he has made his peculiar deposit for substantial and genuine virtue. It is the focus in which he keeps alive that sacred fire, which otherwise might escape from the face of the earth."[1]

In response to Hamilton, many defenders of a farm economy took up the cause of agricultural reform. For the most part, these were well-educated, well-to-do gentlemen who could not have planted an acre of corn if their lives depended on it. But they could see that farming was in trouble. Well aware of recent advances in Europe, they were eager to make American agriculture more productive by developing new tools, methods, and crops. To this end, figures such as John and Samuel Adams, Thomas Jefferson, George Washington, Benjamin Franklin, and John Hancock had tried their hand at experimentation and innovation. At Monticello, Jefferson devoted much of his time to overseeing the growing of tobacco and wheat, as well as fruits and vegetables in his garden, testing hundreds of varieties over the years. He also invented devices such as a seed drill and threshing machine, to make farming more efficient. Reform-minded individuals in several Mid-Atlantic and New England states had come together to promote "scientific" farming, taking Europe as their model.

Shortly before the turn of the century, a founding trustee and vice president of Bowdoin, Samuel Deane, published the *New England Farmer or Georgical Dictionary, containing a Compendious Account of the Ways and Means in which the Important Art of Husbandry, in all its Various Branches, is, or may be, Practiced to the Greatest Advantage in this country.* This volume presaged a flood of similar treatises. In 1794, a Philadelphia group proposed endowing chairs in agricultural studies at the University of Pennsylvania and Dickinson College.[2] Jefferson would urge that the University of Virginia take a similar

---

[1] Jefferson, "Manufactures," *Notes on the State of Virginia*, in Jefferson, *The Writings of Thomas Jefferson*, vol. 4, ed. Andrew A. Lipscomb and Albert E. Berge (Washington DC: Thomas Jefferson Memorial Association, 1907), 290.
[2] Taylor, Agricultural Revolution," 343.

step when it opened its doors in 1819.[1]

What Jefferson and his fellow agrarians had discounted was the increasing attractiveness of Western lands. There, instead of growing all their own food, tending animals, and leading self-sufficient lives, farmers could prosper by producing just one commodity. Migration thus offered a chance to escape a subsistence living. The only way to dissuade young families from moving west was to offer them comparable opportunities in their home states. Small-scale manufacturing seemed the best hope. Expansion of this sector would create jobs and — just as importantly — provide a safeguard against overseas trade restrictions and higher costs. Breaking ranks with strict agrarians like Jefferson, these reformers wanted to see advances in farming *and* manufacturing, convinced that the country's well-being hinged upon being strong in both areas. An early leader in this effort was the New York Society for the Promotion of Agriculture, Arts, and Manufactures, formed in 1791. This organization disseminated information about better ways to make leather, paper, steel, paint, maple syrup, fine wool, glass, and china — methods which would make labor more productive, "render human life more comfortable," and raise the general level of prosperity.[2]

In New England, several "gentlemen farmers" pondered what they could do to make their region more prosperous. For them, too, the success of England's wool industry provided food for thought. In the fifteenth and sixteenth centuries, the demand for cloth had convinced many English farmers to give up growing corn and take up herding sheep. Before the coming of cotton mills, the woolen industry had served as the chief engine of the British economy. It was this successful symbiosis between agriculture and manufacturing which was so appealing to these American reformers: they could preserve a land-based way of life by transforming it into a more profitable and sustainable enterprise. Because of Western crop competition, animal husbandry — raising sheep — now seemed the way to do this. New England's stony pastures and gently sloping hills afforded excellent grazing conditions. Sheep could be kept at a relatively low cost: the investment in labor was much less than in farming. Farmers could count on selling their fleece at nearby markets, since the demand for wool clothing was rapidly increasing: wage-earning artisans, mechanics, shop owners, and carpenters

---

[1] Letter of Jefferson to David Williams, 14 November 1803. *Writings of Thomas Jefferson*, vol. 10, 429-30. Jefferson felt all institutions of higher learning should offer programs in agriculture. This field was, he wrote to Williams, "the first in utility, and ought to be the first in respect."

[2] "Letter of the Agricultural Society to the Friends and Promoters of Rural Economy," *Transactions of the Society Instituted in the State of New York for the Promotion of Agriculture, Arts, and Manufactures*, part 1 (New York: Society for the Promotion of Agriculture, Arts, and Manufactures, 1792), vii-xiii.

could now afford to buy clothing made from it.[1] New England already had a long history of producing wool, on hand-turned, household spinning wheels. Rhode Island, Massachusetts, and Connecticut had plenty of rivers to generate the power needed for mechanized manufacturing.

A leading proponent of a domestic wool industry was an enterprising former merchant from Plymouth, Massachusetts, named Elkanah Watson. During the Revolution, Watson had carried dispatches and funds to Benjamin Franklin in Paris. After the war ended he had stayed on in France, becoming a successful international trader. When he returned to the States, Watson settled in Albany, where he championed such improvements as better street lighting and the construction of canals: a speculator in Western lands, he was among the first to make the case for building a waterway to the West.[2] But agricultural reform became his abiding passion.[3] Applying the knowledge about raising sheep he had acquired in Europe, Watson bred a small flock of Merino sheep (known for their abundant, high-quality wool) and then brought them with him when he moved to Pittsfield, Massachusetts, to take up farming fulltime in 1807.[4]

Watson was convinced that raising sheep could restore prosperity to a part of the country losing population because it was not creating enough new jobs. So was David Humphreys, a hero of the American Revolution, former ambassador to Portugal, and the first president of the Agriculture Society of Connecticut. He had laid the groundwork for this enterprise by importing several dozen Merino ewes and rams from Lisbon to Connecticut, in 1802. At

---

[1] In the early decades of the nineteenth century, many New England farmers took jobs in local manufacturing to "even out the season irregularities" of agriculture. Young men worked in mills to save money so that they could buy their own farms. Clark, *Rural Capitalism*, 97, 101.

[2] J.E.A. Smith, *The History of Pittsfield (Berkshire County), Massachusetts: From the Year 1800 to the Year 1876* (Springfield MA: C.W. Bryan, 1876), 323. Watson's and DeWitt Clinton's competing claims about who had first proposed constructing the Erie Canal turned into a bitter, long-lasting dispute.

[3] Watson was an early member of the New York Society for the Promotion of Agriculture, Manufactures and Art, as well as the New York Society for the Promotion of the Useful Arts.

[4] Watson purchased his sheep from Robert Livingston, a co-founder and first president of the New York Society for the Promotion of Agriculture, Arts, and Manufacturers. Livingston's introduction of a Merino pair in the Lower Hudson Valley in the spring of 1802 was an attempt to reverse its declining agricultural economy. See George J. Lankevich, *River of Dreams: The Hudson Valley in Historic Postcards* (New York: Fordham University Press, 2006), 48. See also Robert R. Livingston, *Essay on Sheep: Their Varieties — Account of the Merinos of Spain, France, &c., Reflections on the Best Method of Treating Them, and Raising a Flock in the United States, Together with Miscellaneous Remarks on Sheep and Woollen Manufactures* (Concord NH: Daniel Cooledge, 1813). Merino wool is considered the finest since it has more fibers — some forty thousand — per square inch of skin than other breeds.

the falls of the Naugatuck River, in Derby, Humphreys built what came to be regarded as the finest woolen mill in the country. A staunch Jeffersonian democrat, he believed the survival of husbandry was essential for "preserving [the nation's] . . . republican character and moral institutions." He also felt that New England could only stem the flow of migration to the West if its farms thrived. In a speech in New Haven, he would declare: "The best means to prevent emigrations, will be to convince our citizens that old and worn land can be renovated and enriched by labour and manure, so as to bear as good crops, as land just cleared of forests . . ."[1]

Outside Pittsfield, Elkanah Watson enlisted a small band of Shakers to breed and take care of his Merino sheep. To call attention to this undertaking, he put a ram and ewe on display under Pittsfield's signature, 126-foot-high Liberty Elm in October 1807.[2] The success of this event inspired Watson to form a Berkshire agricultural society. But when he outlined to a group of local farmers his plan to pursue wool production on the nearby hills, they were less than enthusiastic.[3] They didn't see much to be gained from herding sheep.[4]

This resistance to change would last for years.[5] But Watson was not about to give up. In the fall of 1810, he mounted the first agricultural fair in the United States — an event which attracted hundreds of farmers from neighboring states as well as from within Massachusetts. This novel "cattle show" grew into an annual event of major regional importance and popularity. It was staged with considerable fanfare, including solemn pastoral prayers, celebratory speeches, eye-catching displays of Berkshire manufactured goods, cash awards for the best animals, and a high-noon parade through the city, led by Pittsfield's impressive brass band.[6]

---

[1] David Humphreys, *A Discourse on the Agriculture of the State of Connecticut, and the Means of Making it More Beneficial to the State* (New Haven: T.G. Woodward, 1816), 8-9. For more evidence of Humphreys' desire to see wool manufacturing develop in New England, see his letter to Watson, 18 September 1809. Quoted in Watson, *Memoir*, 343-4.

[2] Smith, *History of Pittsfield*, 324.

[3] Ibid., 327.

[4] Prejudice against herding Merino sheep was common until it became clear that farmers could make a good profit selling their fleece to wool-producing factories. Ezra A. Carman, H.E. Heath, and John Minto, *Special Report on the History and Present Condition of the Sheep Industry of the United States*, Bureau of Animal Husbandry, U.S. Department of Agriculture (Washington: Government Printing Office, 1892), 217.

[5] Even in 1816, "prejudices" against agricultural reforms were still thwarting the introduction of more effective farming methods. The refusal of the Massachusetts legislature to support these efforts was telling. See "Berkshire Agricultural Society," *Independent Chronicle*, 25 January 1816.

[6] Smith, *History of Pittsfield*, 337. Cf. Elkanah Watson, *Men and Times of the Revolution, or The Memoir of Elkanah Watson, including Journals of Travels in Europe and America, from 1777 to 1842*, ed. Winslow C. Watson (New York: Dana,

The success of the Pittsfield fair prompted towns all over the country to mount their own events. The "Berkshire Plan" for devising better ways of raising crops and breeding animals led to the creation of other agricultural advocacy groups. By 1822, Thomas Gold — Watson's successor as president of the Berkshire society — could justly claim that its "fame and influence have extended over the entire surface of the United States; its example followed, its approbation courted, by its extended offspring. It has been recognized, as well as in Europe as in America, as an original, novel plan, and the most excellent organization ever conceived to promote the great interests under its patronage."[1] The key to the region's future well-being and "respectability," Gold told an attentive throng of fairgoers, was its emerging sheep industry. [2]

Watson's banking on imported sheep was soon vindicated. Because of an 1811 trade embargo with England, nearly all imports of woolen goods abruptly ceased.[3] Since few domestic mills were then producing wool,[4] the country faced a severe shortage of clothing.[5] The high prices then commanded by wool made the lowly, temperamental sheep a highly desirable farm animal. Visions of quick profits sparked a Merino "frenzy" in upstate New York and western New England.[6] To make room for expanding herds, tens of thousands of acres were cleared of trees and converted to pasture land, making parts of New England resemble western Ireland. Although the sheep industry would have its shares of booms and busts in the ensuing decades, Elkanah Watson's dream of hilly, stone-walled pastures dotted by sheep was ultimately fulfilled: by the middle of the century Berkshire County was producing roughly half the wool in the United States.

1856), 372-3.

[1] Quoted in Smith, *History of Pittsfield*, 344.

[2] Smith, *History of Pittsfield*, 335-6.

[3] Historically, the British Crown had prohibited the importing of sheep to its North American colonies, as a way of maintaining their economic dependence and perpetuating its own lucrative trade in woolen goods. In the mid-1700s, any colonist caught keeping or trading sheep was subject to having his right hand cut off.

[4] The first wool carding machine started operation in Pittsfield in 1800. The first cloth-producing mill in the county opened a dozen years later.

[5] During the colonial era, few farmers had shown any interest in keeping sheep. Even after the Revolution, the Secretary of the Treasury was told by one Connecticut farmer why this remained unpopular: "In the first place, sheep are apt to be unruly and troublesome, and in the second place they will not pay the expense of keeping equal to other stock." Letter of William Hillhouse to Alexander Hamilton, 6 September 1791. Quoted in Victor S. Clark, *History of Manufactures in the United States, 1607-1860* (Washington DC: Carnegie Institute, 1916), 321.

[6] Some farmers had no prior experience in raising sheep and ended up ruining their farmland through overgrazing. See, for example, J. Lowell, "Rural Economy: Agricultural Report," *Dedham Gazette* (Dedham MA), 1 November 1816.

To process wool into cloth, numerous mills opened up in Massachusetts and Connecticut.[1] But large-scale sheep herding and wool manufacturing did not stop New England's economic decline. Why not? For one thing, the industry was too volatile. Prices soared when there was no foreign competition, but then collapsed once trade with England resumed, after the War of 1812.[2] Furthermore, sheep could be herded more economically in the burgeoning Western states.[3] On top of this, the abundance and relatively low price of Southern cotton, combined with Eli Whitney's time- and labor-saving gin, soon made cloth from this material more desirable than wool, especially in warm climes. Then there was the constitutional conservatism of many Yankee farmers. They saw little point in developing "useful manufactures" because Americans could not hope to match the quality of textiles produced in England.[4]

A final obstacle was New England's mercantile class. Boston merchants did not want domestic wool production to increase because they were already making a good living from importing woolen goods. In essence, they preferred to preserve a colonial economy in the United States — exporting raw materials such as tobacco, cotton, flour, rice, timber, and corn in exchange for foreign manufactured products. Federalist dominance in New England assured that this state of affairs would continue indefinitely. In Massachusetts, even the Republican-controlled legislature would have no truck with proposals to modernize the state's agriculture by concentrating on cash crops or animals. The Berkshire Agricultural Society founded by Elkanah Watson was chronically short on cash, hampering its efforts to spread the word about better methods of breeding and planting. Early in 1812, Watson spent a month in Boston lobbying various members of the state legislature for support, but to no avail.[5]

---

[1] One Berkshire farmer noted that wool from Merino sheep was bringing $187 at market, as compared with only $50 for the fleece of a native animal. See letter from "Another 'Farmer,'" *Berkshire Star*, 11 January 1816. The writer pointed out that this high price was mainly the result of the growth of domestic wool manufacturing.

[2] Carman, et al., *Sheep Industry*, 220-21. Foreign imports, low prices for domestic fleeces, the inferior quality of American wool, and the decimation of New England herds all combined to depress the sheep industry. See "Agricultural: From the 'Bangor Weekly Register,'" *American Advocate and Kennebec Advertiser* (Hallowell ME), 3 May 1817.

[3] Costs for keeping a sheep out West were only about one fourth of what they were back East. The coming of the railroads hastened the end of New England sheep-raising, as cheap Western wool could now be transported eastward at a relatively low cost. Bidwell, "Agricultural Revolution," 690.

[4] See, for example, "Miscellany: From the 'Saratoga Patriot,'" *Columbian Register*, 5 January 1813.

[5] Smith, *History of Pittsfield*, 345. The Massachusetts legislature did agree to charter the new Berkshire organization, however. Gerry had responded to Watson's request for assistance by stating: "I shall promote, in every possible

What was badly needed to boost New England's economy was a dramatic change in international conditions. Trade embargoes and the War of 1812 provided this necessary shock therapy. In its first full year of the conflict — 1813 — goods imported from Europe plummeted by nearly seventy two percent. At the same time, American exports were largely choked off. Southern plantations were cut off from European markets, and prices for their agricultural commodities sank.[1] But New England suffered more: its trade-dependent economy went into a nose dive. Its ports were blockaded, and hundreds of its commercial vessels were seized or destroyed.[2] In light of these consequences, many fervent opponents of domestic manufacturing changed their minds. Some Yankee Federalists who had once opposed this sector because it was inimical to their shipping interests now welcomed it. At the other end of the political spectrum, an elderly Thomas Jefferson conceded that America's "exclusion from the oceans" had forced him to repudiate his long-held agrarian philosophy: "We must now place the manufacturer by the side of the agriculturist," he wrote soon after the war had ended.[3] In this altered situation, domestic producers stepped forward to capitalize on having a captive domestic market — no longer dominated by superior, cheaper European products.[4] After having languished for several

---

way, the preservation and increase of Merino sheep, and the manufacture of woolen cloths." He also agreed with Watson's assessment that a healthy sheep industry in the state would slow migration to the west. Letter of Gerry to Watson, 4 February 1811. Quoted in Watson, *Memoir*, 366. Watson lost this executive backing in Boston when Gerry was defeated in the 1812 gubernatorial election by the Federalist Caleb Strong.

[1] Agricultural interests in both the North and South denounced the embargo with England, with Virginia's John Randolph describing it on the floor of the House as "an attempt to cure corns by cutting off the toes." Quoted in James G. Cusick, *The Other War of 1812: The Patriot War and the American Invasion of Spanish East Florida* (Gainesville: University Press of Florida, 2003), 22.

[2] All told during the conflict, 1,407 merchant ships were seized or burned to the waterline by British men'o'war. The U.S. lost an estimated $8.6 million in customs revenues and $97.4 million in foreign trade profits. Arthur, *How Britain Won the War*, 200.

[3] Jefferson went on to explain: "He, therefore, who is now against domestic manufacture, must be for reducing us either to dependence on that foreign nation, or to be clothed in skins, and to live like wild beasts in dens and caverns. I am not one of these; experience has taught me that manufactures are now as necessary to our independence as to our comfort." Letter of Jefferson to Benjamin Austin, 9 January 1816, *Writings of Thomas Jefferson*, vol. 14, 391. A major editorial voice in favor of manufacturing was the Republican *Niles' Weekly Register*. It regularly published articles on developments in this field.

[4] High labor costs had made it difficult for New England manufacturers to sell at competitive prices. Bidwell, "Rural Economy," 270. In 1810, U.S. cotton makers were turning out fewer than nine hundred thousand yards of cloth, compared to the fifty-three million yards imported from India alone in the years 1807-1808. Miles A. Henry, *Lowell As It Was, and As It Is* (Lowell MA: Powers and Bagley, 1845), 219.

decades, American manufacturing staged a comeback. With the humming mechanical jennies in Moses Brown and Samuel Slater's mills along the Pawtucket River in Providence leading the way, U.S. cotton manufacturing flourished.[1] Twenty mills were added in 1813 and twenty five more the following year.[2] Cotton production in New England grew by an average of sixteen percent annually between 1815 and 1833. It then accounted for two-thirds of the region's manufacturing output.[3] In central Massachusetts, farmers in communities such as Northampton, Amherst, and Hadley quickly switched over to making cotton and woolen goods in home-based workshops. Over six thousand of New Haven's ten thousand-odd residents found jobs in the cotton industry.[4] Production was greatly accelerated when a shrewd textile trader from Newburyport named Francis Cabot Lowell brought back from England (in his head) plans for building power looms and integrating cloth production in a single plant.[5] With the financial backing of several fellow merchants, Lowell established the Boston Manufacturing Company and erected a water-powered mill on the Charles River, in Waltham.[6] There a cadre of newly recruited and trained young ladies was soon churning out thirty miles of cotton cloth each day.[7] By 1816, this mill was generating sales of more than $23,000, and in 1819, just over $124,000.[8] Its owners — the so-called "lords of the loom" — became rich. New England was well on its way to becoming a major "outpost of industrial capitalism."[9]

This manufacturing boom also benefited farmers, who were able to sell some of their produce to workers in the mills. Communities located

---

[1] Bidwell, "Rural Economy," 275. Lawrence Peskin puts the number at 102. Peskin, *Manufacturing Revolution*, 134.

[2] Caroline F. Ware, *The Early New England Cotton Manufacture* (New York: Russell and Russell, 1966), 32. Brown and his partner and son-in-law, William Almy, had profited from having purchased large stores of cotton at low prices before the war commenced.

[3] By 1810, there were ninety-six cotton-producing plants operating in southern New England. Table 1, "New England Cotton Industry Output, 1805-1860," Robert B. Zevin, *The Growth of Manufacturing in Early Nineteenth Century New England* (New York: Arno Press, 1975), 10-13.

[4] Bidwell, "Rural Economy," 281.

[5] It was the introduction of these looms which sparked the famed Luddite protests in northern England in 1811-1812.

[6] John O'Sullivan and Edward F. Keuchel, *American Economic History: From Abundance to Constraints* (New York: Markus Wiener, 1989), 55.

[7] Rosenberg, *Life and Times*, 62, 67, 85, 167, 179, 181, 185, 210-1, 237, 254.

[8] Ware, *Cotton Manufacture*, 70. Production of cotton at Lowell factories rose from 53 million yards in 1840, to 95 million in 1850, and 122 million in 1860. Peak production of 338 million yards was reached just before the end of the century. Information provided at Boott Cotton Mills Museum, Lowell.

[9] Clark, *Rural Capitalism*, 8. However, the impact of manufacturing in terms of employment was very limited. In 1816, only about five thousand New Englanders, or one percent of the labor force, worked in mills or manufacturing plants. Zevin, *Growth of Manufacturing*, 10-11.

---

near cities like Boston prospered from serving this new segment of the population.[1] Farmhands also earned extra cash as blacksmiths, carriage makers, shop owners, barbers, goldsmiths, doctors, lawyers, tailors, printers, and the like.[2] Some towns specialized in manufacturing certain items and became national leaders in their industries. In 1809, Danbury, Connecticut, had several dozen shops making hats; five years later, a single factory was manufacturing twenty four thousand annually, employing a full-time staff of fifty men. These roughly-made fur hats were sold in Europe and as far south as South Carolina and turned Danbury into the preeminent "hatting" center.[3] Other small-scale factories sprang up to meet a demand for locally made products, such as shoes, candles, hemp, and iron tools.[4] In New Haven, Timothy Dwight identified persons making boots, shoes, saddles, bellows, leather gloves, mittens, clocks, hats, cordage, bells, soap, candles, scythes, nails, axes, paper, powder, brushes, chairs, trunks, and straw hats.[5] Mansfield, in eastern Connecticut, came to dominate silk manufacturing not long after the country's first mill built for this purpose opened there in 1810.[6]

But New England's first manufacturing "wave" was short-lived. Once hostilities between the United States and England ceased, and a peace treaty was signed at Ghent, Belgium, in December 1814, the era of protectionism came to an end. Full "liberty of commerce" was proclaimed between the two former belligerents: in the future no goods entering or leaving either nation would be subject to duties. The ink on this treaty had scarcely dried when British merchant ships began dropping anchor in New York and other Eastern ports, heavily laden with cotton cloth and other manufactured goods (most of which had been stored during the war for eventual export). These were sold at laughably low prices in order to recapture the American market and drive domestic competitors out of business.[7] This "dumping"

---

[1] The cluster of towns around Boston — including Salem, Roxbury, Reading, and Lynn — had a population of some eighty five thousand and was considered one of the country's wealthiest areas. "Communication: On the Growth and Decay of Our Cities, Etc.," *Boston Daily Advertiser*, 8 April 1816.

[2] In his statistical survey of New Haven, Timothy Dwight identified a host of small-scale craftsmen and manufacturers, ranging from shoemakers to coopers. The city boasted seven establishments making hats, in addition to forty two grocery stores and forty one selling dry goods. These establishments served the needs of some six thousand residents. Dwight, *A Statistical Account of the City of New-Haven* (New Haven: Walter and Steele, 1811), 40-41.

[3] William H. Francis, *History of the Hatting Trade in Danbury, Conn. from its Commencement in 1780 to the Present Time* (Danbury CT: H.L. Osborne, 1860), 4, 6-7.

[4] Clark, *Rural Capitalism*, 84.

[5] Dwight, *Statistical Account*, 38.

[6] Alfred T. Lilly, *The Silk Industry of the United States from 1766 to 1874* (Boston: John Wilson and Son, 1874), 4. Mansfield would remain dominant domestically until the early 1840s, when its mulberry trees were badly damaged.

[7] In 1816, it cost about a third less to produce cotton in England than it did in the U.S. Letter of Isaac Briggs to Rep. William Lowndes, 12 March 1816, *Niles' Weekly Register* 10: 4 (26 March 1816): 49.

strategy succeeded with a vengeance. Flocking to these cheap imports, American consumers turned their backs on nascent local industries, and many of these saw their profits dwindle and faced "instant destruction."[1] In the cotton industry, only Francis Cabot Lowell's mill at Waltham continued to make a tidy profit, thanks to its streamlined production process.[2] Despite this adverse turn of events, government support for domestic manufacturing remained strong. Legislatures in some states allocated funds to their young factories so that they could survive and expand. Between 1808 and 1816, lawmakers in Albany, for example, voted to extend loans totaling $179,000 to manufacturing firms based in New York.[3]

At the national level, pressure mounted to impose a retaliatory duty, if not outright ban, on imported goods.[4] Petitions poured into congressional offices, imploring the government to protect what President Madison had deemed "this source of national independence and wealth."[5] Representatives of the Southern states came out strongly in favor because they felt that England was now warring on the United States by economic means.[6] It was attempting to perpetuate a colonial dominance through trade policies which were "oppressive, vindictive, and unhallowed."[7] Textile interests — in the North and the South — backed trade restrictions because their livelihoods were at stake: their mills could not long stay open when prices were so "depressed."[8] Some Northern manufacturers added their voices to this chorus

---

[1] Elliott, *Tariff Controversy*, 163.

[2] Cotton manufacturers still using Slater's outmoded method saw their incomes decline sharply. Ware, *Cotton Industry*, 75.

[3] Peskin, *Manufacturing Revolution*, 184.

[4] Some manufacturers wanted a "practical exclusion of foreign goods," as had occurred during the war. Clark, *History of Manufactures*, 274.

[5] Quoted in Elliott, *Tariff Controversy*, 165. Advocates of manufacturing chided opponents for being "contemptible and narrow" in not realizing that the U.S. could rival Britain in this economic endeavor. See "Our Manufactures," *National Standard* (Middlebury VT), 24 January 1816. However, other manufacturers sought to allow some imports to enter the country without duty, so that they could keep their costs down. See remarks of Sen. Jonathan Roberts (Democratic-Republican, Pennsylvania), 15 April 1816, *Journal of the Senate of the United States*, 14th Congress, 1st Session, 464. Roberts presented a memorial from Philadelphia manufacturers, asking that cotton from India be exempt.

[6] Some Southern leaders realized that a healthy manufacturing-based economy in the North would be beneficial to the country as a whole as well as to their own region, since it would thus become a dependable market for cotton and other Southern crops.

[7] "Hancock," "Interesting Miscellany: Our Manufactures," *Independent Chronicle* (Boston), 8 July 1816. During the war, some had argued that the U.S. could force England to sue for peace by threatening to curtail all future imports of woolen goods. See, for example, "Political Miscellany: From the *Boston Patriot*," *American Advocate* (Hallowell ME), 29 January 1814.

[8] Already some forty to sixty thousand workers in cotton-producing plants in New York had lost their jobs due to a sharp drop in demand. Nor were these dire economic straits confined to the textile industry. Many iron workers in

calling for a highly restrictive policy.[1]

One powerful group remained generally opposed to erecting tariff barriers and decidedly out of step with most Northerners. This was Boston's ship-owning and mercantile elite — the same blue-blooded merchants whom Francis Cabot Lowell had originally brought together to bankroll his revolutionary mill in Waltham. In fact, the booming textile enterprise, which rewarded them with dividends averaging fourteen percent annually during the years 1817–1820, was their major source of income.[2] Why would they not want to protect it? Clearly England's unfair trade practices would be harmful to their economic interests. Accordingly, owners of the Boston Manufacturing Company sent Lowell and a board member to Washington in February 1816 to persuade congressional leaders to pass the proposed tariff bill. Their efforts paid off when this legislation — imposing a duty on imports for the first time in American history — was approved by the House by a vote of 88-54.[3] But, overall, these captains of industry preferred to let market forces have free rein. Although, as Federalists, they approved of a strong central government, they did not want to see it involved in regulating or manipulating the economy.[4] For centuries, these wealthy Bostonians had been successful sea traders, counting on ships crisscrossing the Atlantic to bring them opportunity and wealth. Enjoying what amounted to a monopoly, they had grown accustomed to having their way without help from anyone else. Manufacturing was a new enterprise, and some traditionalists viewed it warily. Like many Jeffersonian Republicans, they wanted the country to remain true to its agrarian roots. Manufacturing might have its rightful place in an evolving American economy, but they were not inclined to grant it advantages denied to farmers and men of commerce like themselves.[5]

New Jersey were also now unemployed. Elliott, *Tariff Controversy*, 195.

[1] In October 1815, several owners of cotton-producing factories in and around Providence sent off a petition to Congress, requesting that inexpensive cotton imports from India be prohibited, so that their own enterprises — and the investment made in them — would be protected.

[2] Meyer, *Roots of American Industrialization*, 246. Returns from this mill rose to twenty four percent — four times the average rate — in 1821-1824.

[3] Rosenberg, *Life and Times*, 262-66. The Senate voted in favor later that month. The tariff amounted to a roughly twenty percent surcharge on textiles.

[4] This ideology was evident in an 1815 drawing, published in the Federalist *Columbian Centinel* (Boston), depicting the "Good Ship Massachusetts" safely at anchor and extolling the prosperity that free trade had brought to the nation. For a reproduction of this illustration, see Peskin, *Manufacturing Revolution*, 193.

[5] This argument was made by, among others, Daniel Webster. Some leading Republicans, such as Philadelphia physician and legislator George Logan (a founder of the Pennsylvania Society for the Promotion of Agriculture), accepted that manufacturing had a role to play in the economy, but rejected government assistance to this sector. McCoy, *Elusive Republic*, 223-4.

Leading the Federalist opposition in New England to any interference with free trade was a stern, beetle-browed congressman from New Hampshire named Daniel Webster. An intimidating, silver-tongued orator with a Phi Beta Kappa key from Dartmouth, Webster had made a name for himself by giving voice to New Englanders' outrage over a previous trade embargo and subsequently been elected to Congress, in 1812.[1] In a January 1816 speech before the House an irate Webster argued that the Northern States would never have signed the Constitution and joined the Union if they had known the federal government would one day attempt to exercise such unauthorized power to limit commerce.

What in Webster's humble backwoods background made him such a vociferous spokesman for the rights of wealthy ship owners and bankers in Boston and Portsmouth? He was certainly not "one of them" — at least not in a social sense. But Webster shared with these other New England Federalists an abiding commitment to meritocracy. His conservatism was essentially a defense of privilege and a well-entrenched status quo. Federalists of Webster's generation could not imagine any good outcome from cooperation between the private and public sectors: in economic matters they were inherently at odds. It was this credo which prevented New England Federalists from backing trade restrictions or duties, even if these might revive a depressed regional economy.

Their failure to do so slowed the rise of manufacturing in their states. Whereas their putative political allies — and even Republicans like Jefferson — had come around to accepting that factories were crucial to the nation's economic future, these Federalists refused to help struggling young industries. Instead, they believed manufacturers should rise or fall based on their own merits. Behind this free-market posturing, however, was loyalty to the shipping industry, which had made New England so wealthy and powerful. They did not want to help any other form of economic activity at its expense.[2] As one early twentieth-century historian has put it, they feared any "blow at the flag that floated on the high seas."[3] A strong affinity for England also made the merchant class unwilling to protect fragile American

---

[1] Portsmouth was hard hit by the loss of transatlantic trade, and many people in the state were seething with anger over the ensuing economic collapse. As one biographer of Webster has put it, "Merchants howled their pain as their commerce sank to its knees." Robert Vincent Remini, *Daniel Webster: The Man and His Time* (New York: Norton, 1997), 93. Webster wrote an influential pamphlet contending the federal government's embargo was illegal, as commerce was to be properly regulated by the states, not Washington.

[2] Some Federalist merchants felt that Republicans were advocating for manufacturing in collusion with the French so they could destroy New England's shipping industry. Peskin, *Manufacturing Revolution*, 193.

[3] Roland Ringwalt, "Was Webster Inconsistent on the Tariff?" *The Protectionist* 21:49 (January 1910): 484.

industries by restricting imports.[1] Because of its intransigence in this matter, many factories soon had to shut down, even though they were producing high-quality goods.[2] Federalist control of state legislatures assured that no public funds would rescue them.[3]

The tariff of 1816 gave Northern manufacturers some temporary relief as they struggled to hold their ground against cheap British imports. Still, their revenues did not increase, since many American consumers continued to favor English goods.[4] But domestic companies did gain vital experience in financing, incorporating, staffing, and running mills and factories. Meanwhile, U.S. agricultural exports did not grow either, as England slapped a duty on items coming from overseas. With prices still high, money scarce, bankruptcies increasing, and foreign trade only slowly returning, a mood of "gloom and discouragement" settled over much of New England.[5] No economic sector was doing well. New England's merchant elite waited patiently for shipping to recover. But the curtailing of trade and closing of ports for several years had taken its toll. Cities like Newport, Newburyport, New Bedford, and Salem were mired in recession, with vessels lying idle in their harbors, and young and skilled laborers heading westward to seek their fortune.[6] One Newburyport poetaster summarized the sour mood in these sarcastic 1808 verses:

> Our ships all in motion once whitened the ocean,
> They sailed and returned with a cargo;

---

[1] Sean Wilentz, *The Rise of American Democracy: Jefferson to Lincoln* (New York: Norton, 2005), 183.

[2] Tudor, *Letters on the Eastern States*, 213. Tudor noted that some products, such as sailcloth and linen, were actually superior to their British counterparts.

[3] In Rhode Island, Federalists held majorities in both houses of the legislature during the War of 1812 and afterwards, until 1817. In Connecticut, the party's domination lasted until 1816. Federalists lost control of the state legislature in Massachusetts in 1804, but regained their majority four years later, by increasing the number of delegates from solidly-Federalist Boston. Republicans defeated them again in 1810, ending the era of Federalist power in the state.

[4] Elliott, *Tariff Controversy*, 187. They did so despite the fact that the quality of British imports was often inferior to American-made items. This allowed the drain of U.S. specie abroad — $150 million in 1815 alone — to continue. "Interesting Miscellany," *Independent Chronicle*, 8 July 1816.

[5] "Interesting Miscellany: The Times," *Independent Chronicle*, 19 December 1816.

[6] Newport had been the nation's leading commercial center after the Revolution. But the British blockade and feared invasion during the War of 1812 had driven up prices for commodities so much that working-class families could barely afford to live there anymore. Edward Peterson, *History of Rhode Island* (New York: J.S. Taylor, 1853), 258, 260-1, 271. Between 1810 and 1820, Newburyport's population declined by over ten percent, New Bedford's by thirty percent, and Newport's by over seven percent. Overall, the population of New England increased 8.8 percent during this decade.

Now doomed to decay, they have fallen a prey
To Jefferson — worms — and embargo.[1]

Overseas markets only slowly reopened, and sea-going trade never returned to what it had been in the region's heyday. Before the war a third of foreign imports had entered the United States through ports in Massachusetts, but afterwards cities like Philadelphia, New York, and New Orleans dominated.[2] Salem, whose square-riggers had once charted trade routes across the Pacific to China and the East Indies under the motto *Divitis Indiae usque ad ultimum sinum* ("To the farthest port of the rich East"), returned brimming with tea, spices, silk, porcelain, ivory and gold, and made the city's residents the richest per capita in the nation, now only had fifty or so ships a year docked at its wharves.[3] Families whose centuries-old fortunes had been made from the cargoes carried by these elegant tall-masted schooners saw their bank accounts shrink. Only Boston continued to grow, largely because it absorbed the failing commercial enterprises in nearby seacoast towns and cities.[4] The national locus of economic activity had shifted to the Ohio Valley, where migrants from New England and the Mid-Atlantic states were flocking now that peace had returned.

With New England's commerce in a perilous state, its manufacturing floundering in light of renewed international competition, its political influence in Washington waning because of the Federalists' opposition to the recently ended war, and its population ebbing, the region's outlook was bleak.[5] While the rest of the country was moving forward, New England was lagging behind.[6] If this situation did not improve, emigration would

---

[1] Quoted in Samuel Eliot Morrison, *The Maritime History of Massachusetts, 1783-1860* (Boston: Houghton Mifflin, 1921), 187.
[2] Rosenberg, *Life and Times*, 165. The precise figure was thirty seven percent.
[3] Doug Stewart, "Salem Sets Sail," *Smithsonian Magazine* 25:3 (June 2004): [unpaginated].   http://www.smithsonianmag.com/history-archaeology/salem.html. The proportion of American ships engaged in trade declined sharply from 68.8% in 1810 to a low of 44% in 1822. It never again exceeded 50%. Arthur, *How Britain Won the War*, 105. In his introduction to *The Scarlet Letter*, Nathaniel Hawthorne, a native son (b. 1804) of Salem, would recall the day "when India was a new region, and only Salem knew the way hither." Hawthorne, "The Custom-House, "*Scarlet Letter* (New York: Dodd, Mead, 1948), 35.
[4] Morrison, *Maritime History*, 191.
[5] Between 1790 and 1820, the population center of the U.S. moved 127 miles to the west — from twenty-three miles east of Baltimore to a spot north of Woodstock, Virginia. Jeremy Atack and Fred Bateman, *To Their Own Soil: Agriculture in the Antebellum North* (Ames: Iowa State University Press, 1989), 72.
[6] Even the rise of manufacturing during the war did not offset declines in fishing and commerce. After the conflict ended, the failure of shipping to rebound kept New England's economy in relatively poor shape. Donald R. Hickey, *The War of 1812: A Forgotten Conflict* (Urbana: University of Illinois Press, 1989), 305.

persist and even increase, since British soldiers and Native American tribes no longer posed a threat to frontier settlers. Over time, large swaths of countryside in Vermont, New Hampshire, Massachusetts, and Connecticut would be devoid of inhabitants, man or beast, their once lovingly tilled fields overgrown with weeds and shrubs.[1] Even with the protective tariff working to their advantage, New England's mills and factories were faltering, incapable of providing enough jobs to restore prosperity.[2] The sluggishness in New England's manufacturing sector is apparent in the lack of growth in cities where most factories were located. Only Boston, Hartford, and New Haven experienced any appreciable population increase between 1810 and 1820.[3]

And agriculture was also in a slump.[4] Piles of British goods were depressing the prices for domestic farm produce as well as for New England's manufactured items.[5]

---

[1] As an example of this downward trend, the central Vermont town of Chelsea, founded in 1784, had experienced a population growth of 275 percent between 1790 and 1800, and 47.94 percent the following decade, but only a 10.17 percent increase in the years 1810-1820. Even though growth resumed in the 1820s (up over one third), the town never expanded beyond two thousand residents after then. Its population in 2000 was 1,250. Hal S. Barron, *Those Who Stayed Behind: Rural Society in Nineteenth-Century New England* (Cambridge: Cambridge University Press, 1984), 27.Migration from states like Connecticut did pick up after the war. Richard J. Purcell, *Connecticut in Transition, 1775-1818* (Washington DC: American Historical Association, 1919), 146.

[2] 102 In 1820, only about seventeen thousand people in Connecticut worked in manufacturing, out of a population of 275,000; in Massachusetts the same small proportion — six percent — was employed in industrial plants. These figures are taken from the U.S. census for 1820. In the year ending in September 1816, the U.S. exported $53.3 million in agricultural goods, but only $1.4 million worth of manufactured items. For this statistic, see untitled article, *Weekly Messenger* (Boston), 27 February 1817.

[3] Boston was, by far, the fastest expanding New England city, growing by twenty eight percent during that decade — from 33,787 to 43,298.

[4] Some recent historians have challenged this long-held belief that New England agriculture was faring poorly in the early decades of the nineteenth century. See, for example, David Meyer's *The Roots of American Industrialization*, Winifred Rothenberg's *From Market-Places to a Market Economy*, and Michael Bell's, "Did New England Go Downhill?" *Geographical Review* 79:4 (October 1989): 450-66. Studies like these rely upon data which indicate farmers were making decent profits from the sale of their goods, largely due to greater productivity. However, they do not explain why moving to the Western frontier was still so appealing to such a large number of New England farmers.

[5] This was certainly the case in the Berkshires. See "Miscellany for the Star: To the Farmer," *Berkshire Star*, 11 January 1816. During his 1816 tour through New England, Henry Fearon had noted that prices for produce in Rhode Island and Connecticut were "not commensurate with those of labour." Fearon, *Sketches of America*, 98. The price for wheat in Philadelphia in April 1816 was only eighty-five cents. It would rise to $1.73 the following May, reflecting the impact of poor harvests. John D. Post, *The Last Great Subsistence Crisis in the*

The brief wartime "boom" was clearly over.[1] Wool, for example, was now fetching much less than its previous high of between eight and ten dollars per fleece.[2] Pleas to "buy American" textiles were falling on deaf ears.[3] Prices for wheat, corn, and other crops were also dropping. Rather than bolster one another, agriculture and manufacturing were dragging each other down. The transition to a market economy had stalled. What some had realized before the war was now all too apparent: New England's commercial farming simply could not compete with larger, more productive Midwestern farms.[4] This was particularly true for wheat and beef cattle.

How then could the region jumpstart its economy? A return to subsistence farming would only make matters worse, as it would not create any wealth and only weaken an already fragile urban manufacturing base. A more promising step would be to construct more factories, which would create jobs as well as a demand for regional products. Many business leaders supported this course of action. What was lacking, they asserted, was government funding, to prime the pump and set the idle industrial machinery in motion. Dozens of petitions were mailed to Washington making this case. Painfully aware that their states were in bad economic shape, the governors of several New England and Mid-Atlantic states joined this appeal.

Western World (Baltimore: Johns Hopkins University Press, 1977), 47.

[1] The economic slump was hurting all regions of the country. Thanks to this trade stoppage, the price for Southern cotton — the nation's leading export — had doubled during the war, but now it fell sharply. Murray N. Rothbard, *The Panic of 1819: Reactions and Policies* (New York: Columbia University Press, 1962), 5.

[2] Minto, et al., *Sheep Industry*, 221.

[3] In his October 1817 address to the Berkshire Association for the Promotion of Agriculture and Manufactures, Thomas Gold, a Federalist businessman and banker as well as head of the Berkshire Agricultural Society, appealed to patriotism, saying that it was just as harmful to allow foreigners to make cloth for Americans as it would be to permit them to write our laws. Thomas Gold, *Address of Thomas Gold, Esq., President of the Berkshire Agricultural Society, Delivered Before the Berkshire Association for the Promotion of Agriculture and Manufactures, at Pittsfield, Oct. 2d, 1817* (Pittsfield MA: Phineas Allen, 1817), 28. Cf. "Berkshire Agricultural Society," *Independent Chronicle*, 25 January 1816. The anonymous author of this latter article argued: "We can never say we are, in reality, independent, can never consider our liberties as safe, until we can produce within ourselves, all the comforts and accessaries [sic] of life."One reason that Americans continued to buy British imports is that many of these items were not made in the U.S. For example, American cotton manufacturers produced mainly low-count, or "plain," cotton cloth, whereas the British made a finer material for other purposes. Because the latter required more equipment and larger factories, some Americans preferred not to produce it, as it would replicate England's wholesale industrialization, which they found so distasteful. McCoy, *Elusive Republic*, 224.

[4] After 1820, states of the former Northwest Territory would significantly increase their sales of farm produce in the Eastern states. Prior to that date, only about twelve percent of what was produced on Midwestern farms had been sold outside that region.

New York's popular Democratic-Republican leader, Daniel Tompkins complained that Washington was not doing enough to help struggling manufacturers in his state and implored Albany legislators to approve badly needed subsidies. In Vermont, Jonas Galusha, who had defeated his Federalist brother-in-law to regain the governor's mansion in 1815, declared that the woeful state of manufacturing was causing hardship in the Green Mountain State.[1] He promised to open state coffers to help.[2] The pressure on other state governments to take similar measures was ratcheted up. As far west as Pittsburgh, the local citizenry commissioned a study on the future role of manufacturing, which concluded that the federal government should assume a "sacred guardianship" over this fledgling sector. No country, it was argued, had ever prospered when its wealth derived chiefly from the labor of farmers.[3]

The major obstacle to assisting local manufacturers remained the Federalists, who still controlled many statehouses and governor's offices.[4] Men like New Hampshire's Webster stuck to their guns in reasoning that economies worked best when trade was as unrestrained as possible. "With me," he would declare a few years hence, "it is a fundamental axiom . . . that the great interests of the country are united and inseparable; that agriculture, commerce, and manufactures will prosper together, or languish together; and that all legislation is dangerous which proposes to benefit one of these without looking to consequences which may fall on the others."[5]

In the ensuing months and years, several pivotal developments would cause the New England Federalists to abandon this ingrained opposition and become advocates for American manufacturing. This change of heart would propel New England into the industrial age. The first was the failure of the region's shipping industry and seagoing commerce to bounce back. Although the tonnage of American merchant vessels did increase somewhat, it was England which profited the most from the resurgence in transatlantic trade.[6]

---

[1] Elliott, *Tariff Controversy*, 164, 191-92.
[2] Zadock Thompson, *History of Vermont, Natural, Civil and Statistical.* vol. 2, *Civil History of Vermont* (Burlington VT; Chauncey Goodrich, 1842), 98.
[3] "Domestic Manufactures: Pittsburgh Report," *Hampshire Gazette* (Northampton MA), 4 June 1817. For a full statement of this position, see *Report of the Committee appointed by the citizens of Pittsburgh, at a meeting held at the court-house on the 21st of December, 1816, to inquire into the state of the manufactures in the city and its immediate vicinity* (Washington: [no publisher listed], 1817). By 1816, two thirds of the residents of Pittsburgh were involved with some form of manufacturing.
[4] Early in 1816, New Hampshire, Massachusetts, Rhode Island, and Connecticut had Federalist governors.
[5] Webster, "The Tariff: A Speech delivered in House of Representatives of the United States, on the 1st and 2d of April 1824," *The Great Speeches and Orations of Daniel Webster* (Boston: Little, Brown, 1879), 78.
[6] Gordon Jackson, "New Horizons in Trade," *Glasgow*, vol. 1, *Beginnings to 1830*,

In the 1830s, English shipyards turned out fast clipper ships for voyages to the Far East and steam-driven vessels for Atlantic crossings, giving their nation a distinct edge over the Americans.[1] By the end of the century, half the world's tonnage was being built there.[2] New England shipping merchants lost more money due to new regulations which prevented the owners of commercial vessels from also owning the goods shipped on them. This eclipse of overseas trade forced the wealthy elite to look to the manufacturing sector as its savior.

Entrepreneurs like Francis Cabot Lowell threw their considerable weight behind laws which would help their factories grow and create more wealth. Their success in making textiles profitably during the war with England had given many domestic manufacturers confidence that they could do so again as long as some measure of protection was afforded them. After spending the winter of 1816 buttonholing key Southern legislators regarding the proposed tariff, the Waltham cotton manufacturer directed his attention to changing the mind of New England's most intransigent foe of trade barriers — Daniel Webster.[3] In 1816 Lowell talked the New Hampshire congressman into giving up his seat in and moving to Boston, to work as a lawyer representing the interests of textile manufacturers. Webster's departure from the nation's capital opened the door for more favorable consideration of protectionist measures.

At the other end of the economic scale, democratic-minded farmers and laborers, pleased at the downfall of Boston's shipping oligarchy, regarded manufacturing as a more desirable economic engine, since it offered more New Englanders a chance to have a higher standard of living. "Shall we neglect to improve our national resources to enrich a small class of the community?" asked one editorial writer in Boston's leading Republican newspaper. "Shall we employ the merchant to furnish articles of foreign growth or production, which can be manufactured of a better quality, consequently cheaper, by our own citizens?" Since foreign commerce had long been "honorably protected," why could not the makers of textiles and other commodities now enjoy the same benefit?[4] The rise of the Democratic-Republicans in New England tipped the balance in favor of government subsidies for manufacturers.

---

ed. T.M. Devine and Gordon Jackson (Manchester: Manchester University Press, 1995), 227.

[1] The introduction of iron hulls in the late 1820s and 1830s would give the British shipping industry a further advantage.

[2] *The Columbia Companion to British History*, ed. Juliet Gardiner and Neil Wenborn (New York: Columbia University Press, 1995), 271.

[3] Lowell had previously convinced Webster to make an exception to his free-trade philosophy and support a modification of the tax on foreign cotton imports in 1816. Rosenberg, *Life and Times*, 268.

[4] "From the *Chronicle*: The Election," *Bangor Weekly Register*, 5 April 1817.

In 1818, the Massachusetts legislature — once an impregnable Federalist redoubt, but now controlled by the Democratic-Republicans — voted to commission a study of the state's agriculture and the accomplishments of societies dedicated to its improvement. This report, presented that March, came down firmly on the side of making farming more scientific.[1] It also predicted that the Bay State's future economic health would depend upon the growth of household manufacturing.[2]

Collectively, all of these postwar trends made a compelling case for the United States — and, in particular, New England — to transition from being a nation of farmers to one of factory workers. The hardships caused by economic competition from, and military conflict with, Britain had thrust the young republic into a new era. The opening up of the West had ended New England's centrality, politically as well as economically. Now its well-being required that it embrace new ways of supporting its people. The unusually cold, destructive weather of 1816 made the need for fundamental change acute.

As has been noted earlier, the failed harvests that year pushed countless New England farmers to the brink of bankruptcy.[3] Many were so deeply in debt they had no recourse other than to quit their lands and move inland. The trickle of families out of Connecticut, New Hampshire, Vermont, Massachusetts, and Maine turned into a torrent. No exact figure can be put on the number who left because of these unseasonable frosts and snowstorms, but anecdotal evidence indicates it was much greater than at any other time, before or since.[4] As the packed wagons clogged the dirt roads leading into western New York and beyond, a sense of urgency, if not panic, arose among the remaining residents of New England. The *annus horribilis*

---

[1] This was also the message of wool producers such as David Humphreys. See his *Discourse on Agriculture*, 9.

[2] "Agricultural: Extract from the Report of the Committee in the Massachusetts Legislature on the Agricultural Societies of that State," *Vermont Intelligencer* (Bellows Falls VT), 2 March 1818.

[3] Wheat farmers in western Vermont, for instance, found themselves impoverished after the poor harvest of 1816; some five hundred men who lived in Addison County ended up in court the following year, mostly for unpaid debts. Samuel Swift, *History of the Town of Middlebury, In the County of Addison, Vermont* (Rutland: Charles F. Tuttle, 1971), 94.

[4] Wrote one early nineteenth-century historian of New Hampshire about the "poverty year" of 1816: "Not a few came to the conclusion that it was vain to think of raising their bread on the cold hills of New-Hampshire, and that they must hasten to the remote WEST; where they fondly hoped to find an almost perpetual sunshine and unfailing plenty. — Never was the passion for emigration, then familiarly called the 'Ohio Fever,' at a greater height." John W. Whiton, *Sketches of the History of New-Hampshire, from its Settlement in 1623, to 1833* (Concord NH: Marsh, Capen and Lyon, 1834), 189. In a September 1816 address on Connecticut's agriculture, David Humphreys noted that the exodus from this state in the past twelve months had been "considerable." Humphreys, *Discourse on Agriculture*, 17.

of 1816 cried out for radical steps to be taken. The old way of life, based on Puritan dictates and "steady habits," did not seem viable any more. Some new course of action had to be pursued.

Responding to this discontent was a new generation of political leaders — figures elected as Federalist hegemony had given way to more democratic forces. In Connecticut, the new governor, Oliver Wolcott, Jr., spoke to the state legislature in May 1817 on the westward exodus, lamenting that enthusiasm for it was "excessive" and asking that would-be emigrants reconsider, given the scarcity of schools, churches, roads, and other amenities on the frontier.[1] The governor pledged to do what he could to keep its citizens "contented, industrious, and frugal" and to provide the "most speedy relief" to those forced to choose between living in "poverty" in Connecticut and leaving. As the owner of a cotton mill, Wolcott was inclined to regard textile manufacturing as the best hope for the state's economy.[2] In his inaugural address, he asked legislators to earmark funds to help finance more mills and other manufacturing endeavors.[3] Along with Vermont governor Galusha, Wolcott bemoaned the "ruinous emigrations" of young families, but did little to halt these.[4] He merely authorized a study of this problem.[5] The governor's muted response was criticized by the outspoken Republican editor Hezekiah Niles, who, in essays published in his *Hartford Weekly Times*, blamed Connecticut's "political backwardness" for this outmigration.[6] This publication, along with Niles' more widely read *Weekly Register*, espoused the gospel of manufacturing. Even a staunch proponent of agriculture such as David Humphreys could not gainsay that mills and household textile production had an important role to play in revitalizing the Nutmeg State's economy, even if farming remained "paramount." If manufacturing was languishing, it had to be "re-animated."[7]

The new Federalist governor in Massachusetts, John Brooks, recognized that the days of one-party rule in the Bay State were over. No longer could a candidate from the Boston merchant class take election for granted.

[1] See Wolcott's speech to the Connecticut legislature, published in the *Connecticut Courant*, 20 May 1817, reprinted in *The Peopling of New Connecticut: From the Land of Steady Habits to the Western Reserve*, ed. Richard Buel, Jr. (Middletown CT: Wesleyan University Press, 2011), 102.
[2] His brother, Frederick, had established a mill at Torrington, Connecticut, in 1813.
[3] George ___, "Letter III: From a Graduate to His Father," *Times*, 19 August 1817.
[4] Quoted in Barrows Mussey, "Yankee Chills, Ohio Fever," *New England Quarterly* 22:4 (December 1949): 449.
[5] Wolcott speech, *Connecticut Courant*, 20 May 1817.
[6] Jeffrey L. Pasley, *"Tyranny of Printers": Newspaper Politics in the Early American Republic* (Charlottesville: University of Virginia Press, 2001), 379.
[7] Humphreys, *Discourse on Agriculture*, 17.

Thus, like Wolcott in neighboring Connecticut, Brooks adopted moderate positions which mollified Democratic-Republicans, while papering over his deeply held Federalist beliefs.[1] He acknowledged that the people enjoyed sovereign rights and backed several proposals intended to aid Massachusetts manufacturers.[2] One of these was an 1818 legislative report which recommended that household manufactures and "domestic arts" be subsidized to deter more residents from leaving the state.[3] Similarly, when William Plumer, a Baptist preacher and erstwhile Federalist, was reelected governor in New Hampshire in June 1816, one of his first acts was to endorse a bill granting local manufacturers a tax exemption so that they might attract foreign investors.[4] In a speech to lawmakers a year later, Plumer took note of his state's increasingly unfavorable balance of trade and admitted that agriculture alone could not make New Hampshire economically healthy: its people would have to make their own clothing as well as grow their own food.[5] His Federalist counterpart in Rhode Island, William Jones, also joined with state legislators to bolster manufacturing — not surprising in a state which had been the cradle of American industry.[6]

This commitment of these public officials accelerated New England's manufacturing "revolution." By 1820, the proportion of Americans working in agriculture had fallen to seventy two percent, with the shift to factories taking place mainly in this region.[7] One fifth of the food produced on its farms was being transported to town or city markets for sale. Over the next half a century, New England's urban population grew at a rate twenty times greater than its rural areas: the number of people living in cities soared from

---

[1] Matthew H. Crocker, *The Magic of the Man: Josiah Quincy and the Rise of Mass Politics in Boston, 1800-1830* (Amherst: University of Massachusetts Press, 1999), 63. This bipartisan approach earned Brooks election to seven consecutive terms as governor.

[2] Austin, *History of Massachusetts*, 408. In a July 1816 address to the state legislature, Brooks repudiated the New England Federalist advocacy of states' rights in favor of a strong federal government. Untitled article, *Republican Farmer* (Bridgeport CT), 3 July 1816.

[3] "Agricultural," *Vermont Intelligencer*, 2 March 1818.

[4] William Plumer, Jr., *The Life of William Plumer*, ed. A. Peabody (Boston: Phillips, Sampson, 1857), 435.

[5] Plumer, *Message from His Excellency the Governor of New Hampshire to the Legislature of New-Hampshire, June Session, 1817* (Concord NH: Shaw and Shoemaker, 1817), 8.

[6] The only Federalist ever elected governor in Rhode Island, Jones (a Marine captain during the Revolution) was defeated in his bid for reelection in the fall of 1816, by a mere sixty-eight votes. Thomas W. Bicknell, *The History of the State of Rhode Island and Providence Plantations*, vol. 3 (New York: American Historical Society, 1920), 1130. Rhode Island's major export was cotton, and it led the nation in textile manufacturing until 1815, when Massachusetts surpassed it. William G. McLoughlin, *Rhode Island: A History* (New York: Norton, 1986), 120.

[7] Heidler and Heidler, *Daily Life*, 55.

half a million to 8 million.[1] In the state of Connecticut, manufacturing outside the home had scarcely existed before 1815.[2] Shortly after the introduction of power mills there were 133 cotton and woolen-producing plants in operation. Thanks to the phenomenal growth in cotton cloth production (made possible by Whitney's gin), New England made a giant leap forward in manufacturing. Factories producing woolen goods, other textiles, iron stoves and tools, and machinery proliferated. In 1816, at his small plant in Plymouth, Connecticut, Eli Terry began mass producing inexpensive clocks using interchangeable wooden parts.[3] Inside a workshop in nearby Middletown, a mechanic named Simeon North built the first metal-milling machine, making possible the mass production of pistols and rifles.[4] Around this time, some forty five towns in the Nutmeg State were already producing a variety of manufactured items.[5] Starting in 1816, steamships carried these goods twice each week from New London and New Haven to New York City.[6] Towns in eastern Massachusetts such as Brocton, Lynn, Grafton, and Haverhill pioneered shoe manufacturing: by the 1830s this was the state's second-largest industry. (Hartford and Danbury would dominate Connecticut's shoemaking enterprise.) With the rise of machine production two decades later, Massachusetts' factories would account for forty percent of all the shoes made in the United States. Providence also expanded its industrial base beyond textiles, introducing machine-tool manufacturing and silver making; largely because of these developments, the city more than doubled in size between 1820 and 1840. Overall, the proportion of New Englanders making a living from manufacturing increased from 21 to 30.2 percent during these two decades.[7] The number of persons employed in this sector rose from 17,541 to 27,932 in Connecticut; from 33,464 to 85,166 in Massachusetts; and from 6,091 to 21,271 in Rhode Island.[8] Mainly due to

---

[1] Clarence H. Danhof, *Change in Agriculture: The Northern States, 1820-1870* (Cambridge: Harvard University Press, 1969), 2, 9.

[2] The same was true for Massachusetts. Clark, *Roots of Rural Capitalism*, 95.

[3] Some 3,000 a year were produced. Donald R. Hoke, *Ingenious Yankees: The Rise of the American System of Manufactures in the Private Sector* (New York: Columbia University Press, 1990), 53. The following year, Chauncey Jerome, formerly an employee of Terry's, would begin to manufacture his own, brass clocks at the eventual rate of one thousand per day, selling them initially door-to-door. His company would grow to be the largest clock maker in the world.

[4] David A. Hounshell, *From the American System to Mass Production, 1800-1932: The Development of Manufacturing Technology in the United States* (Baltimore: Johns Hopkins University Press, 1985), 29.

[5] Grace P. Fuller, "An Introduction to the History of Connecticut as a Manufacturing State," *Smith College Studies in History* 1:1 (October 1915): 12-13.

[6] Fuller, "History of Connecticut," 26.

[7] George B. Mangold, *The Labor Argument in the American Protective Tariff Discussion* (Madison: University of Wisconsin, 1906), 78.

[8] These figures are taken from the federal census for 1820 and 1840. It should be noted that the 1840 enumeration included persons involved in

growth in manufacturing, the population of several New England states increased significantly, with Massachusetts setting the pace with a forty percent rise.[1]

Creating new local industries slowed migration out of southern New England, where the factories were concentrated. The advent of large-scale manufacturing made the region again prosperous. The need for labor attracted workers from out-of-state and abroad. It also led to steady increases in the hourly wage –roughly one percent annually up to the start of the Civil War.[2] The income derived from textile mills, leather factories, and whale-oil bottling facilities raised the regional standard of living far above the national average: by 1840, it was twenty five percent higher.[3] This capital generated by manufacturing also made New England and the Mid-Atlantic states much wealthier their Southern brethren, who were still dependent on their cash crops of tobacco and cotton. In their ensuing clash this economic edge would enable the North to prevail.

The miserable weather and inadequate harvests of 1816 did not *cause* New England to become the nation's manufacturing center. The region's evolution from agricultural to industrial production came about as a result of the long-term developments described above –the shortage of fertile land, the decline of farm profits, the opening up of Western states to settlement and cultivation, the resulting population shift to the interior, the demise of overseas trade, and the economic hardships stemming from trade embargoes and war with England. Long before the "year without a summer" these adverse trends had been pushing New England in a new direction. Even those who had steadfastly defended the nation's agrarian tradition and ideals came around to accepting that an economic transformation was necessary. The conversion of Thomas Jefferson came before the summer of 1816. As early as 1809 he had written to Sir James Jay (brother of John) that an "equilibrium of agriculture, manufactures & commerce, is certainly become essential to our independence."[4] But the agricultural crisis caused by these

---

"manufacturing and trades" — not just "manufacturing," as the earlier census had counted.

[1] Boston and Lowell, two centers of manufacturing, experienced the greatest increase: Boston grew from 60,000 to 360,000 in the half a century from 1830 to 1880, while Lowell saw its population increase tenfold — from 6,000 to 60,000. Peter Temin, "The Industrialization of New England, 1830-1880," in *Engines of Enterprise: An Economic History of New England*, ed. Peter Temin (Cambridge: Harvard University Press, 2000), 132.

[2] Ibid., 143. Real wages in New England were higher than those in Europe in the mid-nineteenth century.

[3] Ibid., 143. The Mid-Atlantic states enjoyed a similarly high level of income at that time.

[4] Letter of Jefferson to Jay, 7 April 1809. Quoted in Bradford Perkins, *Prologue to War: England and the United States, 1805-1812* (Berkeley: University of California Press, 1970), 49.

---

unseasonable killing frosts brought matters to a head. The inhospitable weather prompted considerable soul-searching about New England's future. It caused widespread poverty and suffering and forced many thousands of New Englanders to leave. Their departure, in turn, intensified the pressure on state governments to take steps to prop up the sagging economy. Forward-thinking "gentlemen farmers," businessmen, and entrepreneurs considered new ways of earning a profitable living. The risks involved in opening mills and producing manufactured goods seemed more palatable in light of the seemingly dire outlook for local farms. With summer fields covered with ice, their plight was made dramatically clear. The time for a new beginning had come: reluctantly and half-heartedly, many hidebound Yankees now realized that they must adapt to these altered circumstances. If they were going to remain in this cherished corner of the New World, they would have to reshape it to fit their new needs — much as their Puritan ancestors had done when they staked out this dark, forbidding wilderness, raised their axes high, and felled the first trees three centuries before.

But the climate crisis of 1816 had much wider ramifications. It would affect not only how New Englanders responded to the economic challenges of this young century, but also how many thinkers and artists in the Western world — in Europe as well as the United States — perceived human existence and addressed questions about the nature and purpose of life in their works. Highly attuned to changes in the *Zeitgeist*, these philosophical and literary giants –those whom Shelley would proclaim the "unacknowledged legislators of the world," and his heir Ezra Pound describe as the "antennae of the race"[1] — would articulate a truly modern sensibility and craft a bold, highly individualized art, casting aside both the forms and the spirit of their eighteenth-century predecessors. Like Prometheus before them, they would seize the freedom deep in the heart of the godless angst they felt and turn it mightily to their advantage.

---

[1] Shelley, *A Defense of Poetry* (1821) and Pound, *ABC of Reading* (New Haven: Yale University Press, 1934), 73.

Chapter 6. Confronting the Monster Within

"If I cannot inspire love, I will cause fear."
— Mary Shelley, *Frankenstein* (1818)

During 1816, much of northern Europe experienced extraordinary weather as well — torrential downpours, flooding, late snow, and exceptionally cold temperatures. This abrupt change in climate stirred millennial fears and stimulated the imaginations of a generation of writers, artists, and philosophers. This Romantic response to the "year without a summer" then inspired counterparts on the other side of the Atlantic and altered the development of American literature.

When the young couple, their four-month-old son, and their inseparable female companion finally reached Switzerland in early May, 1816, they found no sign of spring — only snow, ice, and freezing cold. It was as if the transition of the seasons had been rudely arrested. On the French border, not far from Lake Geneva, the poet had to hire ten strong men to shovel snow from underneath their carriage so they could proceed further south. The four English travelers huddled under foul-smelling rugs as four horses snorted steam and strained for footing in the deep drifts, and snowflakes bombarded the frosted glass. As night fell, huge black pines rose up beside the road, their boughs sagging with snow, as if in abject defeat. Clutching each other for warmth, the foursome traversed an "icy desert wilderness" in terrifying darkness, with a wild wind rattling the carriage and a snow shower bombarding them like a swarm of maddened white moths.[1] "Never was a scene more desolate," confessed the poet's eighteen-year-

---

[1] Richard Holmes, *Shelley: The Pursuit* (New York: New York Review of Books,

old lover a few days later, in a letter to her half-sister, Fanny. After having gazed upon the warm, tilled fields south of Paris bursting with new life and the promise of more to come after such a long winter and so many years of war, she was depressed by this unseasonal regression. The "tranquility" she hungered for after weeks of endless travel — first across the Channel, then down the Rhine, and finally overland through France — was being cruelly withheld. Nature, too, was conspiring against the fleeing lovers.

Mary Godwin and the man she adored — destined to be the most celebrated literary couple in history — had suffered much on their long winding road to freedom. In setting out on this Continental odyssey, Percy Shelley, twenty-three, had narrowly escaped the clutches of his creditors, much as Mary had eluded her hated stepmother two years earlier by running off with him to France. Spurning the laws of man and God (whose existence they openly denied), the couple had envisioned their return to warmer and more benign climes as bringing a welcome respite from these ongoing struggles, a refuge from the poverty and hardship of battle-scarred France, and a chance to live as they wanted.[1] The southern sun would sooth their consciences, warm their love, and enflame their imaginations. Instead, they found inky tentacles spreading across a pale Swiss sky, as if foreshadowing worse trials to come.

Then, suddenly, the massive clouds parted and the sprawling lake below exposed its dazzling, azure charms as boldly as a guileless lover. They were enchanted. Taking up rooms at Geneva's Hotel d'Angleterre, Mary delighted in the sunbeams flooding the freshly-swept floor and in the droning of insects in the garden — more signs of life returning. Mornings she and Shelley would be awakened by the cooing of tiny Will, reminding the drowsy couple of the happy future they were creating for themselves. During the daytime, Mary and Shelley shed the "gloom of winter and of London" by releasing balloons or sailing on the lake, returning by moonlight to be "saluted by the delightful scent of flowers and newly mown grass, and the chirp of the grasshoppers, and the song of the evening birds." As Mary rhapsodized in one letter, "coming to this delightful spot during this divine weather, I feel as happy as a new-fledged bird, and hardly care what twig I fly to, so that I may try my new-found wings."[2] Even the cloying presence of Claire, her half-sister — who had insisted on coming along to Geneva with mad, selfish dreams of rekindling her doomed relationship with Byron — could not dampen their

---

1974), 322-3. See also Miranda Seymour, *Mary Shelley* (London: John Murray, 2000), 152.

[1] Bill Phillips, "*Frankenstein* and Mary Shelley's 'Wet Ungenial Summer,'" *Atlantis* 28:2 (December 2006): 66.

[2] Letter of Mary Shelley to Fanny Imlay, 17 May 1816, *The Letters of Mary Wollstonecraft Shelley*, vol. 1, *A Part of the Effect* (Baltimore: Johns Hopkins University Press, 1980), 18.

ebullient mood.[1]

Unbeknownst to them and the war-weary people of Europe, the eruption of faraway Tambora the previous year was about to bring a calamity down on their heads, unleashing floods of near biblical proportions. In early June, strong, gusty winds were already rippling Lake Leman, accompanied by sharp, steely rain. Heeding the orders of worried Swiss authorities, Byron and Shelley had to forego their daily sail.[2] Lightning flashes and booming thunderstorms became daily occurrences, glinting and roaring like muffled cannons massed on the opposite shore.[3] Mary described these storms as "grander and more terrific than I have ever seen before."[4] The soothing sunshine gave way to "mist and extreme cold."[5] Temperatures fell, and blankets were unearthed.[6] To escape the brazenly prying eyes of other guests, the Shelleys and Claire moved across the lake to a cottage close to the three-storey Italianate villa Byron had rented, boasting an uninterrupted view of the evening sun nestled over the Jura Mountains. The inclement weather drew them together, spending more time now at their neighbor's commodious abode.[7]

With rain falling nearly daily the leisurely strolls through the gardens had to be called off.[8] Instead, they lost themselves in literary work — Byron was penning the third canto of his *Childe Harold's Pilgrimage*, and Shelley was polishing his "Hymn to Intellectual Beauty" — and lively conversation. (They were joined in this by Byron's young Anglo-Italian physician, John William

---

[1] Claire Clairmont, the stepdaughter of William Godwin (Mary's father), had accompanied Shelley and Mary Godwin when they eloped in 1814. After her advances were rejected by Shelley, she threw herself at Byron in March 1816, bedded him once, and became pregnant with his child before she left for the Continent in the spring of 1816. It was her desire to see Byron again which induced Claire to bring the two poets together. William Dean Brewer, *The Shelley-Byron Conversation* (Gainesville: University Press of Florida, 1994), 7. Many years after Shelley's death, Mary wrote to her friend Edward John Trelawny confessing that Claire "still has the faculty of making me more uncomfortable than any other human being." Quoted in Dorothy and Thomas Hoobler, *The Monsters: Mary Shelley and the Curse of Frankenstein* (Boston: Little, Brown, 2006), 319.
[2] Radu Florescu, *In Search of Frankenstein* (New York: Warner Books, 1975), 156.
[3] Holmes, *Pursuit*, 328.
[4] Letter of Mary Shelley to Fanny Imlay, 1 June 1816, in Mary Shelley, *Selected Letters of Mary Wollstonecraft Shelley*, ed. Betty B. Bennett (Baltimore: Johns Hopkins University Press, 1995), 12.
[5] John Clubbe, "Tempest-toss'd Summer of 1816: Shelley's *Frankenstein*," *Byron Journal* 19 (1991): 27.
[6] It would later be calculated that the mean recorded in Geneva for July 1816 was 4.9 degrees Fahrenheit lower than it had been during the years 1807-1824. Jonathan Bate, *The Song of the Earth* (Cambridge: Harvard University Press, 2000), 96.
[7] 10 The terraced, three-story Villa Diodati had been built in 1710 by a friend of the poet John Milton.
[8] Between April and September 1816, it rained in Geneva 130 out of 183 days. Phillips, "'Wet Ungenial Summer,'" 60.

Polidori, who also happened to be smitten with him.) Being shut in for so long taxed their mercurial dispositions, strained their complex relationships, and darkened their thoughts. The awesome power of the nocturnal storms made them acutely conscious of the "precarious nature of the earth and of life itself."[1] The stormy weather was at once exciting and sobering. In her letters, Mary confessed it left her in a decidedly foul mood.

It was small consolation to find out that people elsewhere were enduring the same miserable summer. Fanny wrote that it had been "dreadfully dreary & rainy" in England since Mary and Shelley had left.[2] In fact, most of northern and central Europe was experiencing its coldest and wettest summer on record — before or since.[3] In Switzerland, as elsewhere, the soaking rain ruined crops and lent fields and meadows a ghastly, wintry look. Lingering alpine snow only added to this dismaying appearance.[4] Melting icepack in the mountains caused more flooding than the Continent had ever witnessed.[5] Lake Geneva rose by seven feet, and neighboring rivers and streams overflowed their banks, in wild, frothing torrents. Bridges were washed away and livestock drowned. Waterways across Europe spilled over their banks: the Seine rose eight feet in a matter of days.[6] Pink snow fell on parts of Italy.[7] The Swiss were hardest hit by the loss of their grain harvest, but other countries also ran short of foodstuffs.[8] Europe faced the greatest threat of starvation since Shakespeare's day.[9] In parts of southwestern Germany and

---

[1] Ibid., 31.

[2] Letter of Fanny Imlay to Mary and Percy Shelley, 29 May 1816, *The Clairmont Correspondence: Letters of Claire Clairmont, Charles Clairmont, and Fanny Imlay Clairmont*, vol. 1, *1808-1834*, ed. Marion Kingston Stocking (Baltimore and London: Johns Hopkins University Press, 1995), 48. The constant rain would make it more difficult for a failing Jane Austen to finish her last novel, *Persuasion*, since her house in Chawton, England, was frequently full of her young nieces and nephews. Claire Tomalin, *Jane Austen: A Life* (New York: Vintage, 1999), 255.

[3] Records kept in Geneva since 1753 indicate no previous summer had as much rain or colder temperatures. Clubbe, "Tempest-toss'd Summer," 29. According to John Post, the temperatures in Europe during the spring and summer of 1819 were "among the lowest in the recorded meteorological history of the Western world." John D. Post, *The Last Great Subsistence Crisis in the Western World* (Baltimore: Johns Hopkins University Press, 1977), 1. The British were experiencing their lowest summer temperatures since 1659. C. Wilson, "Workshop on World Climate in 1816: A Summary and Discussion of Results," in *The Year Without a Summer? World Climate in 1816*, ed. C.R. Harington (Ottawa: Canada Museum of Nature, 1992), 533.

[4] Ibid., 533.

[5] Holmes, *Age of Wonder*, 383.

[6] "Paris, July 16," *American Beacon* (Norfolk VA), 18 September 1816.

[7] Holmes, *Age of Wonder*, 383.

[8] Post, *Subsistence Crisis*, 21.

[9] On the Iberian Peninsula, temperatures two to three degrees Celsius colder than normal prevented dates and grapes from ripening. Respiratory infections usually occurring only in winter were commonplace. Snow was

along the Rhine, signs of famine were ubiquitous: Carl von Clausewitz, who had helped the Prussian army defeat the French at Waterloo the year before, observed "ruined figures, scarcely resembling men, prowling around the fields searching for food among the unharvested and already half rotten potatoes that never grew to maturity."[1] Still recovering from the devastation of the Napoleonic wars, France had no reserves of wheat or other grains on hand, and so prices soared to levels which would not be seen again until the eve of World War I. By the spring of 1817, over eleven percent of Paris's population was impoverished.[2] Bread was also in short supply in Switzerland, and many foresaw mass starvation in the coming months.[3] The doubling and tripling of grain prices put these essential commodities out of reach for large numbers of people in France, Italy, Germany, Belgium, the Netherlands, England, and Switzerland, creating what has been aptly described as the West's "last great subsistence crisis."[4]

This economic disaster had major political and social consequences. (It also inspired some technological innovations: reacting to the exorbitant price of oats — needed as feed for horses — an enterprising German forester named Karl Drais invented a four-wheel contraption for hauling goods by foot power. Dubbed a *Laufmaschine* or velocipede, this was the forerunner of the bicycle.[5]) The extraordinarily cold season had come at a most unpropitious time: war had ended trade among many nations and drained their treasuries. Chilly weather the previous summer had resulted in subpar harvests, so the food supply was already diminished and the populace anxious and on edge.

Riots sprang up like wildfires across the French, Belgian, and British countryside, accompanied by looting and the torching of private and public buildings. There were spontaneous uprisings in Holland, England, Germany, and France. Paris was plagued by lawlessness. There was so much crime in that country that the police were overwhelmed. Arrests in England and

---

seen on peaks in the Montseny range (north of Barcelona), and the Llobregat River in Catalonia was frozen over. Ricardo M. Trigo, et al., "Iberia in 1816, the year without a summer," *International Journal of Climatology* 29:99-115 (2009): 102.

[1] Quoted in Fagan, *Little Ice Age*, 171.

[2] Post, *Subsistence Crisis*, 78.

[3] Clubbe, "Tempest-toss'd Summer," 28.

[4] See John Post's 1977 book with this title. Crop failures were experienced as far east as Russia.

[5] Soon thereafter Drais devised a two-wheel version of this new means of transportation, called a draisine. A newspaper reported in 1817 that its widespread use would likely lead to a sharp drop in the price of oats. John Savino and Marie D. Jones, *Supervolcano: The Catastrophic Event that Changed the Course of Human History* (Franklin Lakes NJ: Career Press, 2007), 71. For more on the connection between this invention and the Tambora eruption, see Mike Hamer, "Histories: Brimstone and Bicycles," *New Scientist* 2484 (25 January 2005). http://www.newscientist.com/article/mg18524841.900-brimstone-and-bicycles.html.

Wales nearly doubled between 1815 and 1817.[1] In Cambridge, a mob grabbed sharpened sticks and swarmed through the streets, yelling "Bread or blood!" Only the summary execution of a several rioters stopped them in their tracks.[2] In eastern Switzerland, mobs went on a rampage, in a desperate search for something to eat.[3] Seeing their governments helpless in this crisis, some Europeans disavowed the liberal politics they had embraced since the defeat of Napoleon. Reports of the deposed French emperor's returning from Elba stirred their hearts.

Millions changed their outlook and plans. Marriage and birth rates fell, as young couples were loath to start families under these unfavorable conditions.[4] Many people in hard-hit regions turned to religion for some hope of surviving the crisis. In Switzerland, Catholic and Protestant churches alike were packed every Sunday. Speculation that the world might be coming to an end rekindled interest in Biblical prophecies and mysticism.[5] Some residents left their countries and set out to find new ones across the Atlantic.[6] The impact of the Tambora eruption was felt as far away as India: unusually cold temperatures, accompanied by heavy monsoon rains, caused widespread flooding and led to an outbreak of cholera during the second half of the year.[7] Slowly, but inexorably, this deadly bacterial infection migrated around the world. By 1831, the pandemic reached Egypt, where it killed twelve percent of the population. From there cholera was brought by Russian soldiers to Poland, and hence on to Hungary and France, where several hundred thousand people succumbed to it. Immigrants carried the disease to New York City the following year: one hundred perished in July 1832 alone. Prolonged dampness in Ireland produced a typhus epidemic between 1817 and 1819, claiming the lives of some forty thousand persons.[8] On the other side of the world — in China — the rain-swollen Yangtze and Yellow Rivers overflowed their banks, and untold numbers drowned.[9] Taken

---

[1] Ibid., 73, 94.

[2] Clive Oppenheimer, "Climatic, Environmental and Human Consequences of the Largest Known Historic Eruption: Tambora Volcano (Indonesia) 1815," *Progress in Physical Geography* 27:2 (June 2003): 251.

[3] Post, *Subsistence Crisis*, 91.

[4] At the same time, mainly as a result of starvation or malnutrition, the number of deaths in Europe rose by nearly ten percent between 1815 and 1817. Ibid., 111.

[5] Ibid., 97.

[6] Some thirteen thousand migrated to the U.S. from England in 1816, another twenty-one thousand in 1817, and twenty-eight thousand in 1818. Roughly twenty thousand Germans made the transatlantic crossing in 1816-17. One study has found that a high percentage of Germans emigrating were motivated by economic factors. Post, *Subsistence Crisis*, 100-1.

[7] Wilson, "Workshop," 538.

[8] Oppenheimer, "Human Consequences," 253.

[9] Jelle Zeilinga de Boer and Donald Theodore Sanders, *Volcanoes in Human*

as a whole, the global impact of the Indonesian eruption made it "one of the very great world disasters associated with climate."[1] None of this, of course, was known to the two handsome, expatriate English poets and their friends and lovers ensconced along the banks of Lake Geneva.

Mary Shelley was the most affected by the powerful storms. Benign, sun-drenched climes afforded her peace and "tranquility," but flashes of lightning on the horizon stirred deeper emotions, including the terror which the philosopher Edmund Burke had associated with the "sublime."[2] But her fascination with the violent thunderstorms over Lake Geneva was alloyed by ambivalence. Mary was torn between desires for a peaceful, harmonious environment and for a volatile, highly charged one. This psychological conflict could be traced to her early experiences. Her childhood — losing her mother at birth, dealing with a hostile stepmother, and living up to the lofty expectations of two famous parents (William Godwin and Mary Wollstonecraft) — had been anything but placid. So Mary longed for unadulterated love and affirmation and the security they could give her. On the other hand, she was her parents' daughter: Godwin had raised her to behave as unorthodoxly as her mother by transgressing the boundaries of social convention and political conviction. Mary had obligingly done so — first by rejecting the tenets of religion, then by adopting radical positions on women's roles, social justice, and individual rights, and finally by running off with a married man notorious for his dubious morals and mounting debts. The haunting tale she would compose that rainswept summer would help her resolve this inner conflict.

During Mary's lifetime, man's relationship with Nature had become problematic. New scientific knowledge had cast doubt on the belief that God controlled all terrestrial events, both extraordinary and ordinary. So how could human beings feel secure? The Deist response was that, since God had created the universe with a benign, rational intent, humans had nothing to fear. Nature was a visible manifestation of the divine will. The bounties, beauty, and harmony of the physical world confirmed that, as Alexander Pope famously expressed it:

History: The Far-Reaching Effects of Major Eruptions (Princeton: Princeton University Press, 2002), 148-51.

[1] H.H. Lamb, *Climate, History and the Modern World* (London: Methuen, 1982), 237.

[2] Clubbe, "Tempest-toss'd Summer," 32. Burke published *A Philosophical Enquiry into the Origin of Our Ideas of the Sublime and the Beautiful* in 1756. His theory that the sublime differed from the beautiful by including sensations of horror or terror had a major influence on early nineteenth-century aesthetics. It is highly likely that Mary Shelley was familiar with this work, as she had read Burke's *A Vindication of a Natural Society* (1756) and *Reflections on the Revolution in France* (1790). See *A Mary Shelley Encyclopedia*, ed. Lucy Morrison and Staci Stone (Westport CT: Greenwood, 2003), 61.

> All are but parts of one stupendous whole,
> Whose body nature is, and God the soul . . .
> All nature is but art, unknown to thee;
> All chance, direction, which thou cannot see.
> All discord, harmony not understood;
> All partial evil, universal good
> And, spite of pride, in erring reason's spite,
> One truth is clear: Whatever is, is right.[1]

However, unmollified doubts about the supremacy of reason gave rise to a new aesthetic sensibility — the Romantic temper. It posited that intense feelings, channeled into art, could convey an *individual* truth more meaningful than the homilies of Neoclassical poets like Pope. English Romantic poets led the way in developing this new sensibility and winning converts to it. The most influential of these was William Wordsworth, whose plainly written but richly evocative verses about Nature gave readers a vivid appreciation for the divine in the ordinary — for what was known as "natural religion."

For Wordsworth, the higher reality glimpsed through the prism of "Nature's pristine majesty" — in a "host of golden daffodils" — was almost always benevolent.[2] Natural beauty was a consolation for worldly woes and distractions as well as a reminder of the ultimate purpose of all living things. In one of his most famous, early poems, Wordsworth thus summarized its restorative power:

> The day is come when I again repose
> Here, under this dark sycamore, and view
> These plots of cottage-ground, these orchard-tufts,
> Which at this season, with their unripe fruits,
> Are clad in one green hue, and lose themselves
> 'Mid groves and copses. Once again I see
> These hedge-rows, hardly hedge-rows, little lines
> Of sportive wood run wild: these pastoral farms
> Green to the very door . . .
>
> These beauteous forms
> Through a long absence, have not been to me
> As is a landscape to a blind man's eye:
> But oft, in lonely rooms, and 'mid the din
> Of towns and cities, I have owed to them
> In hours of weariness, sensations sweet,
> Felt in the blood, and felt along the heart;
> And passing even into my purer mind,

---

[1] Alexander Pope, *Essay on Man* (1732-34).
[2] Wordsworth, "Descriptive Sketches" (1793) and "I Wandered Lonely as a Cloud" (1807).

With tranquil restoration: — feelings too
Of unremembered pleasure: . . .

While with an eye made quiet by the power
Of harmony, and the deep power of joy,
We see into the life of things.
O sylvan Wye! thou wanderer thro' the woods,
Now often has my spirit turned to thee![1]

But not all Romantic poets and writers felt that everything they beheld was "full of blessings." Burke's concept of the sublime contended that the emotions of horror and despair were more powerful than simple delight. A majestically beckoning peak in the Alps, for example, was also cold, remote, and impassive. How could this darker aspect of the natural world be reconciled with belief in a benign universe, even one bereft of God?

A literary style which delved into this fascinating nexus between beauty and terror was the Gothic. Closely allied with Gothic Revival architecture — with its ornate facades, soaring spires, and vaulting arches, as well as its medieval atmosphere of mystery, gloom, and decay — novels and poems in this vein questioned whether rational inquiry and democratic revolutions were actually bringing about some vaunted Golden Age. On the contrary, these developments were both overstepping the proper limits of human endeavor and reducing life to purely material processes.[2] Using tropes of a haunted castle, a damsel in distress, a transgressing Eternal Wanderer, shrouded legends, stormy weather, and supernatural events, the Gothic aesthetic radiated raw, repressed emotion— to jolt the reader into experiencing another, more vital level of reality.[3] These feelings tapped into the subconscious — a deeper probing of human existence than reason alone could make possible.[4] The thrills and chills aroused by these melodramatic tales constituted its visceral appeal as well as its import: the medium was the message.

The combination of "almost perpetual" rainfall which kept the English visitors indoors that fateful summer of 1816, the accompanying, terrifying thunderstorms over Lake Geneva, and the rarefied literary heritage, personal relationships, and emotional state of the teenaged Mary Shelley, created the

---

[1] Wordsworth, "Lines Composed A Few Miles Above Tintern Abbey," 1798.
[2] David Punter and Glennis Byron, *The Gothic* (Malden MA: Blackwell, 2004), 20. See also Jonathan Bate, *Song of the Earth* (Cambridge: Harvard University Press, 2000), 21.
[3] Eino Railo, *Haunted Castle: A Study of the Elements of English Romanticism* (Edinburgh: Edinburgh Press, 1927), 148.
[4] Vijay Mishra, *The Gothic Sublime* (Albany: State University of New York Press, 1994), 19. Cf. Andrew Smith, *Gothic Literature* (Edinburgh: Edinburgh University Press, 2007), 2.

"perfect storm" for her to compose a mesmerizing Gothic tale.[1] Her story of Victor Frankenstein and his "hideous progeny" would not only earn her lasting fame, but also become one of the seminal works in this emerging genre, captivating readers and inspiring imitators for generations to come. Mary Shelley's family background and recent experiences provided the impetus and raw material for this masterpiece. Since her girlhood, William Godwin had exposed his daughter (and only child) to provocative, contemporary thought and advances in a host of fields, including science, in order to arouse her intellectual curiosity.[2] In 1814, the renowned surgeon John Abernethy had presented a series of highly publicized lectures in London, in which he had conjectured that human beings possessed an animating spirit inseparable from the body. This theory had fascinated Mary.[3] Three months before going to Switzerland, she had learned of talks given by a physician and friend of hers and Shelley's — William Lawrence (a former pupil of Abernethy's) — outlining the contrary notion that life could arise without spirit — when some external physical force, such as electricity, was "superadded" to an inert body.[4] Following the practice of her late mother, Mary kept a journal, which gave her a chance to hone her own narrative style and record her observations of — among things — the weather and the countryside.[5] In February 1815, she had given birth two months prematurely to a daughter, who had died less than two weeks later. Even though Mary soon became pregnant again, childbirth would thereafter become an anxiety-producing event for her. Around this time, while still living in England, she and Shelley had read

[1] Letter of Mary Shelley to Fanny Imlay, 1 June 1816, *Letters of Mary Wollstonecraft Shelley*, 20. At least one author has suggested that Mary was also influenced by hearing of the experiments of the German alchemist Johann Konrad Dippel and his Castle Frankenstein while sailing down the Rhine in the spring of 1814. See Florescu, *In Search of Frankenstein*, 95-6. However, Florescu offers no documentary proof that Mary had known about this controversial theologian, physician, and inventor of the dye Prussian blue.

[2] For instance, Mary had gotten to know the chemist Humphry Davy when she was a child. Markham Ellis, "Fictions of Science in Mary Shelley's 'Frankenstein,'" *Sydney Studies in English* 25 (1999): 3.

[3] Mishra, *Gothic Sublime*, 189. However, she sided with Abernethy's rebellious former student, Lawrence, in believing that the body existed wholly independent of the spirit. See Punter and Byron, *Gothic*, 21.

[4] Seymour, *Mary Shelley*, 155.

[5] Mary drew upon her own journal entries and letters from her 1816 trip for her *History of a Six Weeks' Tour through a Part of France, Switzerland, Germany, and Holland, with Letters descriptive of a Sail round the Lake of Geneva, and of the Glaciers of Chamouni* (1817). Mary Wollstonecraft's letters to her former lover, Gilbert Imlay, written during a four-month stay in Scandinavia, had been published shortly before her death under the title *A Short Residence in Sweden, Norway and Denmark*. Ironically, these letters had caused William Godwin to fall in love with Wollstonecraft. Some literary historians have speculated that her daughter wished to achieve the same result (with Shelley) by writing this account of her travels. See, for example, Anne K. Mellor, *Mary Shelley: Her Life, Her Fiction, Her Monsters* (London: Routledge, 1989), 25.

Edmund Spenser's *The Faerie Queen*, in which the rebellious Prometheus fashions a living creature out of parts from animals.[1] In Devon, the two lovers had also visited the laboratory of a scientist, Andrew Crosse, who was well-known for conducting experiments involving electrical currents.[2]

By all surviving accounts (Mary's journal for this period has been lost), several closely related circumstances at Byron's villa led to the writing of *Frankenstein* in June. The first was a series of conversations about the human body and soul which took place around a crackling fire during several lightning-rent evenings. The second was Byron's proposal that the group read aloud German "fantasy" stories, in French translation, to "amuse" each another.[3] The one he mischievously chose to recite on the stormy evening of June sixteenth told of a man kissing his bride on their wedding night, only to find she has morphed into a corpse. To arouse even more fright, Byron read passages from Coleridge's Gothic poem "Christabel" two nights later, at the stroke of midnight.[4] This melodramatic reading achieved its intended effect: Shelley ran from the room, screaming hysterically: he later told Mary he had seen her naked bosom, with eyes staring at him instead of nipples.[5] As the crowning event, Byron suggested that each member of their little literary circle — he, Mary, Percy Shelley, and the physician Polidori — write about some "supernatural occurrence," in a playful literary competition.[6] Mary was at first flummoxed and intimidated by this proposal: in the company of the world's most famous poet and one of its most intellectually complex ones, she felt hopelessly out of her league. Fortuitously, that same night she had a vision of a "pale student of unhallowed arts kneeling beside the thing he had put together . . ." The young man then rushed away from his "odious handywork [*sic*], horror-stricken."[7]

---

[1] Seymour, *Mary Shelley*, 134.

[2] Ibid., 156.

[3] The work from which they read was *Fantasmagoriana, ou Recueil d'Histoires d'Apparitions de Spectres, Revenans, Fantomes, etc.; traduit de l'allemand, par un Amateur* [*Fantasmagoriana, or Collections of the Histories of Apparitions, Spectres, Ghosts, etc.*] (Paris: Lenormant and Schoell, 1812).

[4] Byron had met Coleridge earlier that year and urged him to publish "Christabel," along with other, similar poems (including "Kubla Khan"). This collection appeared in May 1816.

[5] Seymour, *Mary Shelley*, 157. Coleridge's poem contains the following lines, which may have evoked Shelley's dramatic response: Beneath the lamp the lady bowed, / And slowly rolled her eyes around; / Then drawing in her breath aloud, / Like one that shuddered, she unbound / The cincture from beneath her breast: / Her silken robe, and inner vest, / Dropped to her feet, and full in view, / Behold! her bosom and half her side— /A sight to dream of, not to tell! / O shield her! shield sweet Christabel.

[6] Mary Shelley, September 1817 preface to *Frankenstein, or the Modern Prometheus* (London, New York: Oxford University Press, 1969), 14.

[7] Mary Shelley, introduction to 1831 edition of *Frankenstein*, 10. In his journal, Polidori disputes Mary's version of having procrastinated a long time before

The lightning flashes over Lake Geneva merged in her mind with these nocturnal discussions of galvanism to give Mary Shelley the godless "life force" needed to animate this monstrous creation, just as these spectacular atmospherics stimulated her and her companions to pen their "ghost" stories.[1] An excursion into the foothills of the Alps she undertook a month later, on July twenty-first, with Claire and Shelley, intensified Mary's feelings of terror and isolation, as well as her excitement at being in the presence of an overwhelming, indifferent Nature. After crossing through a quiescent "Champagne country," teeming with swelling grapes under blue skies, the trio followed the Arve River to more turbulent and threatening terrain. In the chasm carved out by the roaring current, Mary saw a scene of a "more savage and colossal character." When the hikers reached the village of Chamonix, near the foot of Mont Blanc, she recorded that the waterfall emanating from this locale "fell into the Arve, which dashed against its banks like a wild animal who is furious in constraint." Two days later, as she, Shelley, and Claire continued on their toward the Montanvert glacier, Mary was taken aback by the icy, lifeless, and foreboding landscape: "Nothing can be more desolate than the scents of this mountain; the trees in many places have been torn away by avalanches, and some half leaning over others, intermingled with stones, present the appearance of vast and dreadful desolation." Heavy rain and Shelley's brief fainting spell slowed but did not stop them from reaching their objective — the glacier Mer de Glace. Looking down on it from atop Montanvert, Mary reflected: "This is the most desolate place in the world; iced mountains surround it; no sign of vegetation appears except on the place from which [we] view the scene."[2]

She recalled this vista at several points in her tale of Victor Frankenstein and his "fiend" — in the opening account of Robert Walton's imagining the Arctic as a hyperborean "region of beauty and delight,"; in his encounter with Frankenstein while Walton's ship is trapped amid "plains of ice"; in Victor's return to Geneva and subsequent flight into the Alps; in his search for solitude in the Orkneys with their "appalling and desolate" landscape; and finally in his pursuit of the "fiend" into "the everlasting ices of the north, where — the Monster tells him — 'you will feel the misery of cold and frost, to which I am impassive.'"[3] Such descriptions of the frigid polar realm and the snowcapped Alpine peaks not only heighten the dramatic element in her narrative, but also pass moral judgment on Frankenstein's refusal to accept responsibility

---

starting to write her tale. Seymour, *Mary Shelley*, 157.
[1] Holmes, *Pursuit*, 328.
[2] Mary Shelley, *Journals*, entries for 21, 22, and 25 July 1816, 50-3.
[3] Mary Shelley, *Frankenstein: The 1818 Text, Contexts, Nineteenth-Century Responses, Modern Criticism*, ed. J. Paul Hunter (New York: Norton, 1996), 9, 47, 63, 107, 113, 142.

or show concern for the creature he has brought to life.[1] For these dominions, lacking all human comfort and compassion, symbolize the narcissistic, obsessive nature of Frankenstein's quest for knowledge and creative power. The barren world in which he feels at home stands in stark contrast to the pastoral village life, full of heartfelt compassion, which the fleeing, outcast Monster can only vicariously and briefly enjoy.[2] The underlying premise of Mary Shelley's Gothic story is that God does not exist.[3] An atheist like Shelley and both her parents, Mary rejected any notion of divine control over earthly events. She believed that only the human imagination could infuse the natural world with meaning. Victor Frankenstein's fatal error is to suppose that reason alone can lead to ultimate knowledge. Taking Enlightenment thinking to an extreme, he has reduced existence to logical, comprehensible facets, but in the process lost all sense of its wonder and beauty.[4] Cut off from friends and family and devoid of emotion, Frankenstein undertakes a self-serving, monomaniacal intellectual odyssey which inexorably ends in tragedy: the Monster turns on his creator, seeking revenge for Frankenstein's failure to make him fully human.[5]

The Shelleys' rain-soaked journey southward through France, the weeks of depressing weather along the lake, the violent thunderstorms, and the couple's terror-arousing trek across the glaciers had profoundly affected their sensibilities. The natural world did not always afford "tranquil restoration";[6] at times, it was harsh and dangerous. Human beings could

---

[1] See, for example, George Levine, "The Ambiguous Heritage of *Frankenstein*" in *The Endurance of Frankenstein: Essays on Mary Shelley's Novel*, ed. George Levine and U.C. Knoepflmacher (Berkeley: University of California Press, 1979), 10-11.

[2] While hiding in a hut in this village, the Monster observes that, in this family, "Nothing could exceed the love and respect which the younger cottagers exhibited towards their venerable companion. They performed towards him every little office of affection and duty with gentleness; and he rewarded them by his benevolent smiles." When spring comes, the Creature relates: "The pleasant showers and genial warmth of spring greatly altered the aspect of the earth. . . . The birds sang in more cheerful notes, and the leaves began to bud forth on the trees. Happy, happy earth! fit for habitation for God, which so short a time before, was black, damp, and unwholesome. My spirits were elevated by the enchanting appearance of nature; the past was blotted from my memory, the present was tranquil, and the future gilded by bright rays of hope, and anticipations of joy." *Frankenstein*, 74, 77.

[3] Levine, "Ambiguous Heritage," 7.

[4] Bate, *Song of the Earth*, 54.

[5] The Monster confronts Frankenstein with this pledge: "I will revenge my injuries: if I cannot inspire love, I will cause fear; and chiefly towards you my arch-enemy, because my creator, do I swear inextinguishable hatred." Shelley, *Frankenstein*, 99. The fact that his creation is emotionally "botched" suggests that science cannot fully accomplish its stated ends. See James Rieger, *The Mutiny Within: The Heresies of Percy Bysshe Shelley* (New York: George Braziller, 1967), 88.

[6] Wordsworth, "Tintern Abbey."

not count on divine intervention to ward off its ravages. Being alone in the universe incited fear — the source of terror, as Burke had postulated. Having witnessed the destructive force of the weather in 1816, the Shelleys had to reexamine their previous philosophical outlook.[1] Uncertainty, vulnerability, and existential angst would hereafter be incorporated into what they would write.

In *Frankenstein* Mary would elevate these hyper-natural experiences in Switzerland into manifestations of a state of being — an example of what John Ruskin would disparagingly label the "pathetic fallacy."[2] The almost palpable chill one feels in reading passages describing these climes vividly conveys their emotional coldness. As an antidote, Mary presents the modest virtues of ordinary domestic life.[3] In this sense, her novel is an implicit indictment of self-absorbed Romantic excess — and of the poet most associated with this — Shelley himself.[4] For his part, Shelley saw isolation — financial, political, social, artistic — not so much as a choice but as a given for his ethereal soul. Recurrent pain in his lungs and a doctor's prediction in 1815 that he was dying from consumption had only made Shelley more self-preoccupied. However, recovering his health that summer in Devon, in southwest England, he had found solace in communion with a salubrious Nature:

> Earth, Ocean, Air, beloved brotherhood!
> If our great Mother has imbued my soul
> With aught of natural piety to feel
> Your love, and recompense the boon with mine;
> If dewy morn, and odorous noon, and even,
> With sunset and its gorgeous ministers,

[1] One literary critic has noted that *Frankenstein* is a "reflection of . . . concerns at a time when the natural world was in crisis." The Monster "symbolizes the capacity of nature to instigate environmental crises of biblical proportions." Phillips, "'Wet, Ungenial Summer,'" 59.

[2] John Ruskin, "Of the Pathetic Fallacy," *Works of John* Ruskin, vol. 2 (New York: John Wiley and Sons, 1885), 152-6. In the twentieth century, some literary critics developed the closely related notion of an "objective correlative." For an influential discussion of this concept, see T.S. Eliot, "Hamlet and His Problems," *The Sacred Wood: Essays on Poetry and Criticism* (New York: Knopf, 1921), 7.

[3] See Andrew Griffin, "Fire and Ice in *Frankenstein*" in *Endurance of Frankenstein*, 51.

[4] See Anne Bernays, preface, Joan K. Nichols, *Mary Shelley: Frankenstein's Creator: First Science Fiction Writer* (Berkeley: Conari Books, 1998), xii. Bernays observes: "Into her story Mary wove, both consciously and — we must assume — unconsciously, the travails of her life and that of her family, as well as the wounds she suffered at the hands of the two men she loved most — her father and Shelley . . .." In *Frankenstein*, male figures are self-contained and go off on pursuits, while females are left to nurture and sustain others." Kate Ellis, "Monsters in the Garden: Mary Shelley and the Bourgeois Family" in *Endurance of* Frankenstein, 124.

And solemn midnight's tingling silentness . . .
. . . have been dear to me;
If no bright bird, insect, or gentle beast
I consciously have injured, but still loved
And cherished these my kindred; then forgive
This boast, beloved brethren, and withdraw
No portion of your wonted favor now![1]

But his tempestuous sojourn in France and Switzerland the following year would sorely test Shelley's "natural piety" toward the world around him. This change is evident in the two major poems he composed then. The first is the "Hymn to Intellectual Beauty." According to Mary, this was conceived during Shelley's sailing outings with Byron on Lake Geneva, when rain squalls and fierce winds abruptly arose and thwarted their plans. A dense, highly complex poem, the "Hymn" pays homage to the "awful shadow of some Unseen Power" — a "Spirit of Beauty" which seems to allude to poetic inspiration. But this occasional visitor, "unknown awful," cannot always be counted upon to work its magic. Shelley explicitly compares this fickle and ephemeral "intellectual beauty" to changes in the weather, coming to

This various world with as inconstant wing
As summer winds that creep from flower to flower;
Like moonbeams that behind some piny mountain shower,
It visits with inconstant glance
Each human heart and countenance.

When this spirit departs, it leaves the human imagination deprived of nourishment, as when the sun ceases to shine on midsummer fields. The darkness gives rise to feelings of "gloom," in a world now "vacant and desolate." In the final stanza of "Hymn to Intellectual Beauty," the poet looks past the vagaries of summer to the autumn, when "harmony" will return: he hopes the same "calm" will descend upon his "passive youth" and engender greater creativity, as "solemn and serene" fall days yield the year's final harvest.[2] Acutely aware of his dependence upon congenial external circumstances to conjure up rich poetic images and associations, Shelley can only wish for a favorable environment — and suffer when the "Spirit" remains absent or passes away. It is evident that the extremely variable weather in Switzerland had moved him greatly, causing him to wonder if he could always count upon inspiration. In a world prone to "inconstancy," subject to the "boundless realm of unending change," Shelley was now keenly conscious that little could be taken for granted.[3] This realization reinforced his belief that every moment had to be lived to the fullest.

---

[1] Percy Shelley, "Alastor, or The Spirit of Solitude," 1815 (published 1816).
[2] Shelley, "Hymn to Intellectual Beauty" (1816).
[3] Shelley, "On Death" (written 1814–1816, published 1816).

In short, the sporadic storms and bursts of lightning during the summer of 1816 strengthened Shelley's "Romantic temper" — a temperament which, shorn of any "connections of the finite with the infinite, of the senses with the ideal in the supernatural," stressed the "great importance of spontaneity, passion, individuality, and sought a new kind of "transcendence" in beauty.[1] In reaction to the Neoclassicism of writers like Pope, Romanticism put the highest premium on intensity, not harmony. Those experiences which gave rise to the most powerful emotions were valued the most. While disturbing, these sensations were to be sought out for their inspirational quality. The motifs common to Gothic literature — the forbidding castle, decaying monastery, gloomy forest, impassible mountains, recurring ghost, endangered maiden, melancholy, ominous stranger, and stormy weather [2] — served as the wellsprings of Romantic creativity much as the figure of the Madonna inspired devoutly Christian artists of the medieval and Renaissance periods.

In "Mont Blanc," Shelley's other significant poem crafted at this time, he also explores the fear-evoking quality of Nature, as it was revealed to him on a bridge spanning the Arve. Shelley would describe these verses as conveying the "immediate impression of the deep and powerful feelings excited" by contemplating the massive, snow-draped peak — as an "undisciplined overflowing of the soul . . . an attempt to imitate the untameable [*sic*] wilderness and inaccessible solemnity from which those feelings sprang."[3] Here, too, the inconstancy of Nature is what first strikes the poet — at one moment, the mountain is "glittering," in another "reflecting gloom." The ravine though which the Arve flows can likewise change from a "many-colored, many-voiced vale," sparkling with sunlight, into an "awful scene," with the mighty river current racing down from the "ice-gulfs" guarding Mont Blanc's "secret throne" with a roar like a "flame of lightning." More than its protean nature, this Alpine peak — at 15,774 feet, Europe's highest — fascinated Shelley because of its otherworldliness. Its giddy height seemed to set it apart, put it in touch with the "infinite sky," and elevate it to a level of being which defies human understanding — "Remote, serene, and inaccessible." While, far below, mortal creatures are subject to the ravages of time and earthly turmoil, Mont Blanc "gleams on high," untouched by all such calamities, mute, but all-knowing:

> The secret strength of things,
> Which govern thought, and to the infinite dome
> Of heaven is as a law, inhabits thee![4]

---

[1] Benedetto, Croce, *History of Europe in the Nineteenth Century*, trans. Henry Furst (New York: Harcourt, Brace and World, 1963), 42-7. Poems appearing in Shelley's first publication — "Original Poetry by Victor and Cazire" (1810) — had shown evidence of his penchant for "terror-romanticism." Railo, *Haunted Castle*, 148.
[2] Railo, *Haunted Castle*, 148.
[3] Shelley, preface to *History of a Six Week Tour*, vi.
[4] Shelley, "Mont Blanc: Lines Written in the Vale of Chamouni" (1816).

The higher power Shelley perceives in this majestic mountain is indifferent to human suffering. It offers none of the mollifying grace or forgiveness of a Christian God — none of the comfort Wordsworth found in a field of daffodils. It remains amoral, randomly dispensing good and evil, like a "careless and death-dealing tyrant."[1] All that the mind — and the poetic imagination — can gain by contemplating an apparition like this is to expand the capacity to comprehend the world as it truly is. Mont Blanc teases out inquiring thoughts, but remains itself as silent and enigmatic as the Sphinx. Shelley's response to this peak sets him apart from Romantic poets like William Blake and Wordsworth, who saw in Nature the fingerprints of a supernatural entity.

The two other male occupants of the Villa Diodati that summer also created works which attest to the impact of its strange, portentous weather on their states of mind. The better known one was Byron. It was he who had dreamed up the ghost-story "competition" as a diversion and then provoked his housebound companions by reciting Coleridge's "Christabel." But he, too, was moved by Nature's spectacular nightly displays.[2] Like Shelley, Byron had previously shown a susceptibility to the Gothic temperament, notably in his 1807 collection *Hours of Idleness* and *The Giaour*, published six years later.[3] During one of their June sails on Lake Geneva, Byron and Shelley had come ashore to visit the dungeon where a sixteenth-century monk and Swiss patriot had been imprisoned, at the Chateau de Chillon. Their brief stop had inspired Byron to compose a sonnet about this pitiful figure, which he later expanded into *The Prisoner of Chillon*. Its depiction of endless confinement evoked the prolonged rainy weather which had detained the two poets at a hotel after their sail:

> Lake Leman lies by Chillon's walls:
> A thousand feet in depth below
> Its massy waters meet and flow;
> Thus much the fathom-line was sent
> From Chillon's snow-white battlement,
> Which round about the wave inthrals;

---

[1] Leighton, *Shelley and the Sublime*, 60.

[2] For this competition Byron composed a fragmentary poem entitled "The Vampyre," which was later incorporated into his longer poetic work, "Mazeppa." Coincidentally, in the year Byron had been born — 1788 — his birthplace, London, had experienced extremely cold temperatures as the result of a volcanic eruption on Iceland five years before. Falling ash and poisonous gases had killed several thousand persons in Europe and caused many crops to fail. This consequence, in turn, had ignited popular unrest, building support for the French Revolution. On Iceland itself, a quarter of the population died of starvation.

[3] The *Hours of Idleness*, published when Byron was nineteen, contains poems about a visit to a dead beloved's grave and the decay of Byron's ancestral Newstead Abbey.

> A double dungeon wall and wave
> Have made — and like a living grave
> Below the surface of the lake
> The dark vault lies wherein we lay.[1]

During the remaining days of June spent at the Villa Diodati, Byron was writing the third canto of his lengthy, overtly autobiographical poem, *Childe Harold's Pilgrimage*. In it he recounts the wandering of a heroic alter ego, whose "soul . . . will not dwell/In its own narrow being, but aspire/Beyond the fitting medium of desire;/And, but once kindled, quenchless evermore,/ Preys upon high adventure . . ." Seeking a respite from "the wild world I dwelt in," Harold arrives at Lake Geneva (Leman). There he finds the "hush of night," scent of flowers, and chirping insects Mary Shelley had noted in her letters. But this peaceful interlude does not last. One night Harold sees that:

> The sky is changed! — and such a change!
> Oh night,
> And storm, and darkness, ye are wondrous strong,
> Yet lovely in your strength, as is the light
> Of a dark eye in woman! Far along,
> From peak to peak, the rattling crags among
> Leaps the live thunder![2]

At first, like Mary Shelley, Byron's persona thrills at this display of Nature's extraordinary power: it enflames his yearnings for struggle and conquest. But seeing these summer storms release their fury with utter disregard for the well-being of earthly creatures also caused Byron to reconsider his belief that the forces governing the universe were benign. This newfound skepticism found its voice in a dramatic poem, *Manfred*, which Byron began to compose a month or so after Mary Shelley had started to write *Frankenstein*. These verses expressed the same sense of an aloof, cruel power — embodied by Mont Blanc — which was evident in Shelley's eponymous poem and Mary's novel. Byron even goes farther in presenting himself as a defiant rebel against its might, albeit in vain:

> Ye avalanches, whom a breath draws down
> In mountainous o'erwhelming, come and crush me![3]

As the dreary summer progressed, the periodic thunderstorms, constant rainfall, and gloomy skies took on a more sinister dimension in the poet's mind. His mood was also darkened by recent unhappy events in his personal life — the end of his marriage, the resulting separation from his daughter, Ada, and rumors about his alleged incestuous relationship with his half-

---

[1] Byron, "The Prisoner of Chillon" (1816).
[2] Byron, Canto III, "Childe Harold's Pilgrimage."
[3] "Manfred: A Dramatic Poem," Act I, Scene II (1816-1817).

sister, Augusta Leigh.[1] These developments had caused him to wonder if he shouldn't "blow his brains out."[2] No doubt fully aware that similarly apocalyptic weather was afflicting much of Europe, Byron allowed his imagination to speculate about what this strange change in climate might bode. He was not the only one to do so.

As was noted above, many Europeans interpreted this exceptional weather as a sign that the end of the world was at hand.[3] Early in June, a Neapolitan priest named Carillo had informed his congregation that the planet's destruction would take place on the twenty seventh of that month.[4] An astronomer in Bologna had predicted that all life on earth would perish after the extinguishing of the sun on July eighteenth.[5] The previous sighting of large and numerous sunspots — thought to be harbingers of solar demise — seemed to adumbrate this dreadful outcome. Word of the astronomer's prophecy spread from Italy to France, given greater credence by the incessant rain and ruined crops: "Alarm and consternation pervaded all ranks . . . the churches were filled with devotees, and the event was waited with patient dread by even persons ashamed of openly avowing it."[6] Many townspeople in western Austria gathered in the streets on July fourteenth, trembling with apprehension. Hearing the blare of trumpets during a thunderstorm, more than half the people in Ghent rushed out of their homes, shrieking and moaning, surmising these notes emanated from an angelic "seventh trumpet" heralding the end of time as foretold in Revelation. (In fact, the piercing brass notes came from a passing cavalry regiment.[7]) Newspapers carried reports of persons rioting, committing suicide, and praying fervently for salvation. A large white cloud in the shape of a mountain hovered over the then-Dutch city of Liege, causing considerable panic.[8] Devout Christians,

---

[1] Geoff Payne, *Dark Imaginings: Ideology and Darkness in the Poetry of Lord Byron* (Bern: Lang, 2008), 33.

[2] Byron, *Byron's Letters and Journals*, vol. 5, *"So Late into the Night,"* 1816-1817, ed. Leslie A. Marchand (Cambridge: Harvard University Press, 1976), 165.

[3] During the nineteenth century, perhaps in reaction to a growing secular trend, several years — 1816, 1832, 1857, and 1861 — were marked by an upsurge in such "doomsday" thinking. See Eugen Weber, *Apocalypses: Prophecies, Cults and Millennial Beliefs through the Ages* (Cambridge: Harvard University Press, 1999), 120.

[4] "Paris, July 16," *Enquirer* (Richmond VA), 14 September 1816.

[5] Amedee Guillemin, *The World of Comets*, trans. and ed. James Glaisher (London: S. Low, Marston, Searle and Rivington, 1877), 460. Cf. Jeffrey Vail, "'The Bright Sun was Extinguis'ed': The Bologna Prophecy and Byron's 'Darkness,'" Wordsworth Circle 28:3 (Summer 1997): 183.

[6] William Jones, *Credulities Past and Present; including the Sea and Seamen, Miners, Amulets and Talismans, Rings, Word and Letter Divination, Numbers, Trials, Exorcising and Blessing of Animals, Birds, Eggs, and Luck* (London: Chatto and Windus, 1880), 286.

[7] "Courtray, July 12," *Boston Intelligencer*, 14 September 1816.

[8] Vail, "Bright Sun," 186. Significantly, apocalyptic fears did not take hold

well knowing what the Bible prophesied about the Day of Judgment, the Second Coming of Christ, and the End Time, believed these events were now nigh. Because it was generally believed that the earth was only six thousand years old, people in Europe and the United States could not take comfort in a concept of time stretching over billions of years; it was easier for them to imagine that humanity's end might be imminent.[1] Many biblical interpreters concurred that the "last days" described in Scripture were due to commence 180 years prior to the end of the Christian era — by their reckoning, in 1820.[2] The Kingdom of God would ensue after a period of turmoil and "tribulation," much like what Europe had just suffered during the Napoleonic Wars.[3]

With rivers flooding all around and the sky above nightly ablaze, Lake Geneva was virtually the epicenter of this impending calamity and speculation about it. Some were inclined to regard the spate of storms as proof that the world was about to end — even after the dire July eighteenth prediction was proven wrong.[4] Such eschatological speculation mirrored Byron's own gloomy mood.[5] One summer evening, following a day "on which the fowls went to roost at noon, and the candles were lighted as at midnight," he sat down at a desk in his rented villa and wrote the first lines of what would become his poem "Darkness":[6]

---

across the Atlantic. American newspapers were inclined to scoff at these. A New Hampshire paper referred in September to "silly reports" and "ridiculous apprehensions" in Europe about the imminent end of the world. Untitled article, *The American* (Hanover NH), 4 September 1816. This latter article pointed out that clusters of sunspots had been observed since the late seventeenth century, but no correlation between these unusual phenomena and the weather on earth had been established.

[1] The first discovery of fossilized dinosaur remains — by the English geologist William Buckland — in a cave in Yorkshire, indicating life had existed on earth much longer than six thousand years, did not take place until the early 1820s.

[2] This date was later corrected to 1816, based upon a recalculation of the year of Christ's birth.

[3] See Matthew 24: 3-29, in which Jesus discusses the circumstances of his Second Coming. For a discussion of relevant biblical interpretation circa 1816, see W. Coldwell, "Europe in the Winter of 1830," *The Imperial Magazine and Monthly Record of Religious, Philosophical, Historical, Biographical, Topographical, and General Knowledge,* vol. 1 (London: Fisher, Son, and Jackson, 1831), 83.

[4] Byron's close friend John Hobhouse wrote in his diary on that date: "The day the world was to be at an end." Vail, "Bright Sun," 187.

[5] It is not clear that Byron was aware at that time of the doomsday prophecies which were then appearing in French and Swiss newspapers, but this seems likely since he did keep up with European events during his stay on Lake Geneva. On July eighteenth, Byron noted that he suffered from a mood of depression brought on by "stupid mists — fogs — rains — perpetual density." See Vail, "Bright Sun," 185, and Byron, *Letters and Journals,* 86.

[6] Byron referred to the circumstances of his composing "Darkness" in a letter to his good friend (and Shelley's cousin) Thomas Medwin. Ernest J. Lovell, *Captain Medwin, Friend of Byron and Shelley* (London: Macdonald, 1963), 299. The poem was written between 21 July and 25 August 1816.

> I had a dream, which was not all a dream.
> The bright sun was extinguish'd, and the stars
> Did wander darkling in the eternal space,
> Rayless and pathless, and the icy earth
> Swung blind and blackening in the moonless air . . .

Rich in biblical allusion to the Second Coming, this poem laid out a chilling, Bosch-like vision of the world's final days, when cities and forests alike would be consumed in flames, wars would break out all over, famine would spread, the earth would freeze, and men would forget "their passions in the dread/Of this their desolation; and all hearts/Were chill'd into a selfish prayer for light."[1] The despair resonating in these verses captured both the *Zeitgeist* in Europe that summer and Byron's own state of mind. Like the natural world described in "Darkness," he was "emptied of spirit" because of his personal travails and their impact on his creativity. Just as Shelley had found no supernatural compassion when staring up at the snow expanses of Mont Blanc, so did Byron discern no "inherent moral order" in the universe.[2] Humanity faced doom and destruction in an impassive void. On a personal level, thoughts of the same fate haunted Byron, as he would make more explicit when completing his dramatic poem *Manfred* the next year:

> Abbot. This should have been a noble creature: he
> Hath all the energy which would have made
> A goodly frame of glorious elements,
> Had they been wisely mingled; as it is
> It is an awful chaos — light and darkness,
> And mind and dust, and passions and pure thoughts
> Mix'd, and contending without end or order, —
> All dormant or destructive: he will perish,
> And yet he must not . . .[3]

"Darkness" lends a chilling voice to the millennial paranoia which had taken hold in Western Europe during the summer of 1816.[4] Because of Byron's notoriety, the poem attracted a large audience, and thus the emotions it expressed were greatly amplified, like a wail inside an echo chamber. Unlike Shelley, Byron was enormously popular during his lifetime.

---

[1] Byron, "Darkness" (1816). Critics have noted that Shelley's telling Byron of his idea of a world "freezing into extinction" may have provided additional inspiration for this poem. William D. Brewer, *The Shelley-Byron Conversation* (Gainesville: University Press of Florida, 1994), 60.
[2] Ronald A. Schroeder, "Byron's 'Darkness' and the Romantic Dis-Spiriting of Nature," *Approaches to Teaching Byron's Poetry*, ed. Frederick W. Shilstone (New York: Modern Language Association of America, 1991), 114.
[3] Byron, "Manfred," Act III, Scene I.
[4] Jeffrey Vail has noted that the poem speaks to a "shared experience" of the poet and his audience that summer. Vail, "Bright Sun," 184.

When his volume containing "Darkness" — entitled *The Prisoner of Chillon, and Other Poems* — was published toward the end of the year, some seven thousand copies were bought in the first week alone — twenty five hundred more than the first cantos of *Childe Harold's Pilgrimage* sold in six months in 1812.[1] In Byron, the strange weather of 1816 thus found its most eloquent and influential interpreter. Within his verses was the germ of a "modern" self-awareness. Whereas Enlightenment thinkers and writers had dethroned an unpredictable God and replaced Him with immutable laws, the poets and philosophers of the second half of the nineteenth century would question this framework and seek true meaning within the self. The poet was his own god — a realization at once exhilarating and frightening.

But there was a third writer in that lakeside villa whose response to the gloomy atmosphere that summer left an even more enduring cultural mark. John Polidori, hobbled by a sprained ankle, had remained in the villa with Mary while Byron and Shelley were tacking across the lake and, during "two idle mornings," produced his own "ghost" tale. It would be published a few years later as *The Vampyre*.[2] This was the first English-language story about this legendary, blood-sucking creature, and it was a huge success, inspiring hundreds of imitators and attracting legions of readers.[3] (The slender volume was initially given more notoriety by appearing under Byron's name.) Such eminent writers as Edgar Allan Poe, Nikolai Gogol, Sheridan Le Fanu, Alexandre Dumas, and Leo Tolstoy would eventually try their hands at writing "vampire" stories, as the public's salacious appetite for them refused to slacken. This subgenre of horror fiction would reach its apotheosis in 1897, when Bram Stoker brought out his novel *Dracula*.

While Polidori's original story — set as it was in a warmer clime — did not describe the cold, inclement weather then prevalent in Switzerland, it did show unmistakable signs of its influence:

> Twilight, in this southern climates, is almost unknown; immediately the sun sets, night begins; and ere he had advanced far, the power of the storm was above — its echoing thunders had scarcely

---

[1] When Byron's poem *The Corsair* was published in London in February 1814, it broke that record, with first-day sales of ten thousand copies — the entire printing.

[2] Letter of John Polidori to Henry Colburn, 2 April 1819, in Polidori, *The Diary of Dr. John William Polidori* (London: E. Matthews, 1911), 15. Polidori modeled the libidinous main character, Lord Ruthven, on his employer — a form of homage which did not endear him to Byron. His protagonist is described as possessing "that high romantic feeling of honour and candour, which daily ruins so many milliners' apprentices." Polidori, *The Vampyre: A Tale* (London: Sherwood, Neely, and Jones, 1819), 30.

[3] In mid-June Byron had written a "fragment of a novel" about a mysterious, cold, and evil aristocrat traveling through the Middle East, and Polidori was influenced by this.

an interval of rest — its thick heavy rain forced its way through the canopying foliage, whilst the blue forked lightning seemed to fall and radiate at his very feet. Suddenly his horse took fright, and he was carried with dreadful rapidity through the entangled forest. The animal at last, through fatigue, stopped, and he found, by the glare of lightning, that he was in the neighbourhood of a hovel that hardly lifted itself up from the masses of dead leaves and brushwood which surrounded it. Dismounting, he approached, hoping to find some one to guide him to the town, or at least trusting to obtain shelter from the pelting of the storm. As he approached, the thunders, for a moment silent, allowed him to hear the dreadful shrieks of a woman mingling with the stifled, exultant mockery of a laugh, continued in one almost unbroken sound . . .[1]

These sensual, dramatic works by Polidori, Byron, and the Shelleys breathed new life into the Gothic aesthetic in Europe. While this had been present for nearly half a century (starting with Horace Walpole's 1764 *The Castle of Otranto* and continuing with poets like Coleridge and novelists like Ann Radcliffe[2]), it had not yet become greatly popular. In fact, novels and stories showing a predilection for blood, violence, and horror had produced more ridicule than admiration. Jane Austen's *Northanger Abbey* (1817) was perhaps the best-known contemporary parody of the Gothic style.[3] Such spoofs grew out of the prevailing forward-looking, optimistic, and rational outlook on both sides of the Atlantic during most of the nineteenth century. The Gothic critique of the "modern" age remained very much a minority point of view. It would take the eclipse of middle-class Victorian mores and the rise of a more open-minded outlook at the turn of the century for works like Robert Louis Stevenson's *Strange Case of Doctor Jekyll and Mr. Hyde*, Conan Doyle's "The Hound of the Baskervilles," Oscar Wilde's *The Picture of Dorian Gray*, Stoker's *Dracula*, and Henry James's "The Turn of the Screw" to capture the public imagination. Eventually, the Gothic works written on the shores of Lake Geneva in 1816 would find their own mass audience, in addition to engendering a host of novelistic (as well as cinematic) progeny. To this day, Frankenstein's Monster coming to life amid the crackle and

---

[1] Polidori, *Vampyre*, 45-6.

[2] Radcliffe (1764-1823), the daughter of a London haberdasher, wrote some of the earliest Gothic novels in English. Her titles include *The Mysteries of Udolpho, The Romance of the Forest*, and *Gaston de Blondeville*.

[3] "The storm still raged, and various were the noises, more terrific even than the wind, which struck at intervals on her startled ear. The very curtains of her bed seemed at one moment in motion, and at another the lock on her door was agitated, as if by the attempt of someone to enter. Hollow murmurs seemed to creep along the gallery; and more than once her blood was chilled by the sound of distant moans." Jane Austen, *Northanger Abbey* (London: Bentley, 1848), 140. Austen was directing her parody at Ann Radcliffe's *The Mysteries of Udolpho*.

sizzle of electrical discharges in a storm-swept castle laboratory remains one of the most fascinating images in popular culture.

In the years directly after the publication of *Frankenstein*, other writers — British and American — tackled similar Gothic themes and storylines.[1] Coleridge had already shown a penchant for these — in poems like "Frost at Midnight, "Dejection: An Ode," "Ballad of the Dark Ladie," and "Three Graves," as well as in his reviews and other critical writings.[2] In 1817 he published a volume of verse, *Sibylline Leaves*, containing several earlier works — "The Raven," "Ode on the Departing Year," and "The Rime of the Ancient Mariner" — all written in this melancholy vein. Although Byron and Shelley moved on to other interests — satire (*Don Juan*) and the struggle for Greek independence for the former, political freedom (*The Revolt of Islam* and *Prometheus Unbound*) for the latter — some contemporaries continued to mine the fertile literary territory they had explored during the "year without a summer."

One of these was Washington Irving. After moving to England in 1815, in an effort to save his family's commercial business from bankruptcy, the celebrated author of *A History of New-York from the Beginning of the World to the End of the Dutch Dynasty* began looking around for new subject matter to turn into lucrative prose. Encouraged by Sir Walter Scott (whom he visited in the summer of 1817), Irving took to writing fables with a darkly humorous bent. Several of these were published serially in *The Sketch Book of Geoffrey Crayon, Gent.*, starting in 1819. "Rip Van Winkle" was followed by "The Legend of Sleepy Hollow," and Irving overnight discovered he was famous — the toast of literary circles in London and Paris. That Irving had read Mary Shelley's *Frankenstein* is beyond question. Even though Mary's name did not appear on the cover until the revised 1831 edition, her identity as the author was well known. The extent to which the American was influenced by this dazzling work by the fourteen-year-younger poet's widow is more difficult to ascertain. One can readily find Gothic motifs in "Rip Van Winkle" and "The Legend of Sleepy Hollow," but that does not necessarily mean that

---

[1] Ash in the skies over England that summer may well have affected at least one painting subsequently rendered by J.M.W. Turner — namely, "Chichester Canal" (1829).

[2] In 1808, Coleridge had written in his notebook: "The vicious taste of our modern Radcliffe, Monk Lewis, German Romances—take as a specimen the last, I have read, the *Bravo of Venice* in the combinations of the highest sensation, wonder produced by supernatural power, without the means— thus gratifying our instinct of free-will that would fain be emancipated from the thraldom [sic] of ordinary nature—& and would indeed annihilate both space & time—with the lowest of all human scarce-human faculties—viz— Cunning—Trap door—picklocks—low confederacies &c. Can these things be admired without a bad effect on the mind." Coleridge, *Notebooks of Samuel Taylor Coleridge*, vol. 3, ed. Kathleen Coburn (Princeton: Princeton University Press, 1990), 3449.

Irving borrowed them from *Frankenstein*. Mary's having been stimulated by German "stories of ghosts" she had read in Geneva may have prompted Irving to consult these sources when he toured the Continent in 1821 looking for suitable tales for his own purposes.[1] What is known is that Irving and Shelley met in London, in July of 1824, and developed a close, quasi-romantic friendship. Initiated by Mary, this relationship did not blossom into love, and Irving remained a bachelor the rest of his life.

Other major American authors of the first half of the nineteenth century developed a fondness for Gothic tales after having read Mary Shelley, Percy Shelley, Byron, and novelists like Ann Radcliffe and Matthew Gregory Lewis. Their darkly brooding, obsessed characters reappeared, in slightly different guises, in the poetry and fiction of Henry Wadsworth Longfellow. This influence is most transparent in poems like "The Skeleton in Armor," "Autumnal Nightfall," "The Reaper and the Flowers," "Dirge over a Nameless Grave," and "Memories." During an 1836 European journey, Longfellow visited Lausanne and soaked up its medieval beauty, embodied by the city's towering, six-hundred-year-old Gothic-style cathedral. "There are tombs, with marble figures lying upon them, whose features have been hardly less effaced by time than the mouldering forms in the vaults beneath," he jotted in his journal. Reading the inscription on the tomb of an Englishman, Longfellow found his eyes welling up with tears. "A Gothic cathedral is always a work of wonder and delight. The Gothic style is to me the most beautiful."[2] Longfellow's classmate at Bowdoin College and lifelong friend, Nathaniel Hawthorne, wrote even more extensively in this richly Romantic style, in novels like *The House of the Seven Gables* and *The Scarlet Letter*, as well as in short stories such as "Young Goodman Brown."

Edgar Allan Poe, who had spent the summer of 1816 as a seven-year-old boarder at a school in London, was also influenced by the English Romantic poets, particularly Shelley, Coleridge, and Keats. He admired the Platonic idealism of Shelley's poems, as well as the purity and "utter abandonment" in the "The Sensitive Plant."[3] Poe read widely in Gothic literature and borrowed themes and situations from works like *Frankenstein* in crafting his own "horror" stories — "The Masque of the Red Death," "The Murders in the Rue Morgue," "The Cask of Amontillado," "The Pit and the Pendulum" — in

---

[1] Irving was more drawn to these richly imaginative fantasies than to the "realistic" historical fiction then in fashion back in the United States. The principal writer in this style was James Fenimore Cooper.
[2] William Wadsworth Longfellow, *Life of William Wadsworth Longfellow: With Extracts from His Journals and Correspondence*, vol. 1, ed. Samuel Longfellow (Boston: Ticknor, 1886), 230.
[3] See Poe's (unsigned) review of Joseph R. Drake's *The Culprit Fay* and Fitz-Greene Halleck's *Alnwick Castle* in *Southern Literary Messenger* 2:2 (April 1836): 332.

the 1840s. In the following decade, the monomaniacal Victor Frankenstein served Herman Melville as one of the models for Captain Ahab in *Moby-Dick* — a novel he conceived of soon after reading Mary Shelley's tale.[1] Moreover, Melville's search for deeper, metaphysical meaning in Ahab's pursuit of the White Whale echoes Frankenstein's portrayal of human estrangement from Nature and God and the ensuing torment.[2]

***

The appeal of Gothic literary works during the nineteenth century (and beyond) attests to misgivings about the European-American belief in inevitable social, political, and economic progress. It also reveals unease with Enlightenment trust in natural law and in a well-ordered, predictable, and benevolent universe. The Gothic exploration of deep, shadowy emotions aroused by the uncanny and other terrifying phenomena laid bare an aspect of human reality not explicable by Newtonian theorems. Several startling scientific discoveries and concomitant speculations would further chip away at the trust which Neoclassical thinkers had placed in a mechanistically functioning universe. Existing "laws" could not be reconciled with these theories and breakthroughs — Louis Agassiz's theorizing, in 1837, that the earth had experienced several ice ages in the distant past; John Phillips's mapping of geological time over eons, in 1841, based on fossil records; Darwin's postulating, in the 1840s, an "evolution of species" driven not by divine design but by intra-species competition; Karl Marx's declaring, in his 1843 work on Hegel, that religion was merely a "false consciousness"; Friedrich Nietzsche's affirming, in his 1882 work *The Gay Science*, that God was dead; or Sigmund Freud's revealing, in *The Interpretation of Dreams* (1899), how unconscious drives account for much of human behavior. The century which had begun with a questioning of religious truths ended with a debunking of "rational" ones. In this latter assault on accepted wisdom,

---

[1] Mary R. Reichardt, introduction, *Moby-Dick*, ed. Mary R. Reichardt, (San Francisco: Ignatius Press, 2011), xi. Melville was given a copy of *Frankenstein* in 1850, by his friend Richard Bentley. It had an immediate, powerful impact on his imagination. Andrew Delbanco, *Melville: His World and Work* (New York: Knopf, 2005), 129.

[2] As biographer Delbanco has written, Melville adumbrated the "angst of modern life" in *Moby-Dick* and his other novels. Much like Mont Blanc — another majestic white "monster" — the whale symbolizes the void at the center of the universe. Delbanco, *Melville*, 13, 252, 281. Other literary critics have pointed out that it is this "ultimate meaninglessness" against which Ahab rails. See, for example, Robert Alexander, "Apocalyptic Readings of *Moby-Dick*: What Ishmael Returns to Tell Us," in *Moby-Dick*, 669. For a discussion of the parallels between Shelley's response to the Swiss peak and Ahab's fascination with Moby Dick, see Shawn Thomson, *The Romantic Architecture of Herman Melville's Moby-Dick* (Cranford NJ: Associated University Presses, 2001), 12. Unlike the Shelleys, however, Melville never could abandon his own search for evidence of God in the world He had created. Delbanco, *Melville*, 13, 252, 281.

the writings of Romantic and — more specifically — Gothic novelists, playwrights, short-story writers, and poets served as a vanguard. And, in this company, Percy and Mary Shelley, Lord Byron, and John Polidori — shaken by what they witnessed on a storm-ravaged Lake Geneva in June and July of 1816 — stood in the forefront.

Just how much this extraordinary spell of bad weather affected the *Weltanschauung* of nineteenth-century European and American artists and thinkers is difficult to say. Literary influences are usually too subtle, too indirect, and too unconscious to lend themselves to definitive analysis. Some writers have acknowledged the impact of works by their contemporaries or predecessors, and other influences can be imputed from the books in an author's library, or those referred to in correspondence. We can also infer indebtedness by detecting parallels in subject matter or style. But any other search for antecedents eventually turns into informed speculation. Recently, thanks to the electronic scanning of texts, scholars have at their fingertips a more objective tool for plotting the arc of influence over decades and centuries.[1] They can calibrate the frequency with which telltale words or phrases appear — for instance, how often a term like "desolation" is used in works published in English between 1810 and 1850. How frequently a word was used may not explain *why* it was used, but the time period when usage becomes more common limns the historical setting in which this term gained greater cultural currency. Such correlations can thus broadly trace a "history of ideas." The appearance, prevalence, and disappearance of new words or phrases — for example, "cognitive dissonance" — indicate when novel forms of thought or expression took hold, when they flourished, and when they ceased having much relevance.

If one looks for words linked with the "Romantic temper," several patterns emerge. In American English usage (as well as in English fiction), Romantic (or Gothic) terms such as "imagination," "passion," "infinite," "secret," "desolation," "despair," "gloom," "chasm," "yearning," "mysterious," "mutable," "awful," "feeling," "void," "profound," "lonely," "nothingness," and "sublime" appeared in print more often in the first two decades of the nineteenth century than previously, then the frequency leveled off, and thereafter these words were used to roughly the same degree until the second half of the century.[2] Starting around 1810 and continuing to the end of the 1880s, other "Romantic" terms — "love," "terror," "fear," "barren," "death," "veil," and "horror" — become less common. These findings — while not consistent — suggest that reference to intense emotions and moods

---

[1] http://books.google.com/ngrams.
[2] However, words and phrases associated with either the Enlightenment or orthodox Christianity, such as "reason," "universal good," "happiness," "God's displeasure," and "divine Providence," also conform to this pattern.

came into literary vogue after the American and French revolutions — an era marked by the flowering of individual rights and the weakening of political and religious authorities. An inclination to use language obliquely challenging the established order can be glimpsed more than a decade *before* the end of the Napoleonic wars, the War of 1812 — and the climate crisis of 1816. In other words, these seminal events did not bring about a new outlook, but reinforced an existing tendency to believe that human beings could — and should — take charge of their own destiny. But this sense of empowerment came with negative psychological consequences. Being free and self-determining also made individuals vulnerable — unprotected against the vagaries of natural forces. For many Europeans and Americans, a reassuring certainty about the cycles of earthly renewal had been shaken. If God's guiding hand did not oversee their fate, individuals had to forge their own raison d'être and sense of purpose. This was a daunting, frightening undertaking. The burden of justifying one's existence was heavy; the chances of doing so uncertain. But, nonetheless, the modern era was unfolding, and despite the anxiety it might arouse, there could be no turning back. From now on, each man and woman would have to fashion a Frankenstein in their own image.

POSTSCRIPT

A year later, it seemed that the "year without a summer" had never happened. The weather in New England reverted to its normally volatile, but not calamitous, ways. The preachers and the politicians prayed for a bountiful harvest, and they got their wish. In Montpelier, Governor Jonas Galusha, who had nearly drowned as a young boy and then, as a Green Mountain boy, survived bursts of grapeshot from General Burgoyne's redcoats, had good reason to believe God had smiled on him once again:[1] after asking his fellow Vermonters, in the early spring of 1817, "to entreat the blessings of Him, without whose aid all the efforts of man prove abortive," and to offer up "humble supplications for his pardoning Love and restoring goodness," they were rewarded with sunny skies, warm temperatures, and ample rainfall.[2] At Providence, Rhode Island, the temperature during the first week of July averaged seventy one degrees, up from fifty eight the year before.[3] In the District of Maine, abundant rain and sunshine augured the best growing season in living memory. People counted themselves fortunate and stopped talking about moving west. As an editorial in the *Bangor Weekly Register* that July reassured: "The beneficent hand of providence has hitherto so bountifully dealt out its sun-shine and showers as to preclude every pretence [*sic*] for a murmur

---

[1] Pliny H. White, *Jonas Galusha: The Fifth Governor of Vermont – A Memoir* (Montpelier VT: E.P. Walton, 1866), 4-5.

[2] Jonas Galusha, "Annual Fast," *American Yeoman* (Brattleboro VT), 11 March 1817. Connecticut's Federalist governor John Cotton Smith had likewise declared a day of prayer and fasting that April, urging all persons to express contrition to God for their "abuse of His mercies, our repeated violations of His law, and our inattention to the gracious invitations of His gospel . . ." Proclamation of John Cotton Smith, *Register* (Windham CT), 13 March 1817.

[3] "The Weather," *Providence Patriot*, 12 July 1817.

among the sons of labor."[1] Across the Atlantic as well, more propitious weather had returned, as heavily-laden ships arriving from London reported in mid-summer.[2]

Elsewhere, all was not going so well. Out of western New York in June came news of foul weather eerily reminiscent of what had occurred there twelve months before, awakening "serious apprehensions and alarm in the mind of every one capable of reflection." Late frosts were destroying garden vegetables, six inches of snow was blanketing the fields, and tender saplings in the forest were turning "black as sackcloth."[3]

Overall, the return of favorable weather dispelled the lugubrious mood of 1816. Doomsayers who had foretold an irreversible change in the climate — and in God's attitude toward His people — now offered encomiums to divine Providence. The unsettling fluctuations of the previous year were not cause for alarm, they said, but fell well within the "circle of universal order." Those who had despaired were now chastised for harboring such misgivings. Chided one correspondent in the Brattleboro *Reporter*: "There is no reason to think the chain of physical events is broken, or the tide of human affairs diverted from its ancient channel, because we, who see *through a glass darkly*, observe some apparent irregularities in the seasons." Rather than ponder the causes of these meteorological aberrations, readers should trust in God's constancy: "As the Supreme Being is infinite in wisdom and unchangeable in all his laws . . . we should look upon every appearance in the terrestrial worlds [sic] with an eye of faith, and rest ever satisfied and delighted with this regulation of the Universe."[4] Some disparaged the recent frenzy to migrate as "an alarming disease" which thankfully had now subsided. Prudent New Englanders had come to their senses.[5] (In fact, many families were still making the arduous trek to the Ohio Valley.[6] As one newspaper

[1] "The Season," *Bangor Weekly Register*, 12 July 1817.
[2] "From England," *Providence Patriot*, 23 August 1817.
[3] "The Season," *The Star and North-Carolina State Journal*, (Raleigh NC), 22 June 1817. Datelined "Canandaigua, NY," this article was reprinted from the *Savannah Republic*, 7 June 1817.
[4] "Phocion," "Spots on the Sun," *Reporter* (Brattleboro VT), 6 May 1817. This article had been originally published in the *Boston Gazette*.
[5] Anonymous, untitled article, *Dartmouth Gazette* (Hanover NH), 4 June 1817. Cf. "Emigration," *Dedham Gazette* (Dedham NH), 25 July 1817. The latter piece berated young families for succumbing to this hysteria: "It seems to have been so ordered, that the restless and distrustful spirits 'who travel from Dan to Beersheba, and cry -- all is barren,' might learn that no region of our country is exempt from occasional privations, and none doomed to continued cold and scarcity."
[6] "Ohio Fever," *Bangor Weekly Register*, 5 July 1817. Reprinted from the *Keene Sentinel* (Keene NH). Cf. "Emigration," *New York Post Herald*, 7 October 1817. Reprinted from the *Springfield Federalist*. This article pointed out that the coming of fall was prompting a new exodus.

reported from Cincinnati, the influx of settlers had "rarely been as great as it was this spring."[1])

And, indeed, the 1817 harvest did turn out to be so plentiful that farmers from northern Vermont to southern Connecticut mopped their brows and smiled with relief. "A fruitful and promising season has dissipated the boding fears of the timid, and rewarded the efforts of the enterprising," asserted one Massachusetts editorialist. New England farmers now realized that the West was not some new "paradise" and that they were better off staying where they were.[2] No region of the country was without its "miseries of life," and those who had foolishly imagined otherwise and preferred to be — as another newspaper put it — "the first in an obscure village, than the second man in Rome" now had plenty of reason to regret their hasty, ill-conceived departure.[3]

On the political front, it was a time of flux and new configuration. In some states, the Federalists lost ground to the upstart Republicans, but elsewhere they held on to power. In New Hampshire, the state's final Federalist governor, John Taylor Gilman, stepped down in June 1816 after fourteen years in office. But a month before, the Federalist gubernatorial candidate in Massachusetts, John Brooks (who had led the Middlesex militia in suppressing Shays' Rebellion) had won his race. However, Brooks would be the last major elected official to belong to this party. He would remain in office until 1823. In Connecticut, John Cotton Smith had turned back his challenger, Oliver Wolcott, Jr., in April 1816, but the Republicans had gone on to score an historic victory when Wolcott won the rematch the following May. It seemed that a new balance of power was emerging. After the long, bitter divisiveness associated with the War of 1812, the two major political parties — and the voting public — longed for compromise and accommodation, as was reflected in the Connecticut outcome and Wolcott's policies. At the national level, this desire for peace and tranquility showed in the election of James Monroe to the presidency. A Virginia Republican and onetime law student of Jefferson's, Monroe promised to end partisan bickering and extend a hand to the weakened (and fast fading) Federalists in an "Era of Good Feelings." During the summer and fall of 1817, the new president made an extensive tour through New England, stopping in New

---

[1] "Singular Arrival," *Newburyport Herald*, 1 August 1817. This issue reprinted an article datelined "Cincinnati, July 4." The migration of New Englanders to Ohio, Illinois, and Indiana continued into the fall. Innkeepers reported that the number of persons passing through western Massachusetts was "far greater" than in past years. See *Gazette of Maine* (Portland), 14 October 1817. This article was reprinted from the *Connecticut Courant*.

[2] "Emigration," *Salem Gazette*, 16 September 1817. Reprinted from the *Dedham Gazette*.

[3] "For the Yeoman," *American Yeoman* (Brattleboro VT), 23 September 1817.

Haven, Hartford, New London, Springfield, Newport, Newburyport, and Boston, before going on to Maine.[1] Arriving in the Massachusetts capital on horseback, at the head of a mile-long procession, Monroe was cheered by some forty thousand enthusiastic onlookers. In the course of his Boston sojourn over the Fourth of July, the President broke bread with superannuated John and Abigail Adams and then sat down to mend fences with former Federalist foes such as Harrison Gray Otis. The charm offensive of this "plain, grave-looking man, of dignified deportment, and thoughtful visage" was highly effective[2]: political enemies who had "once yearned to eat Monroe alive now loved him . . . or at any rate lavished their very best upon him whether of wit, wine, or compliments."[3] This visit did much to advance the cause of national unity and mollify the "dangerous prejudices" which had kept the Federalists and Republicans at each other's throats for so long.[4] Clearly, the political climate in New England was undergoing a transition. Republicans appeared to be in ascendancy, but it was too early to say that they would hold on to power for as long the Federalists had.

In religious matters, a similar transformation was taking place. Protestant sects like the Methodists and Baptists were gaining the right to practice their faiths and establish churches in many parts of New England. In 1817, a visitor from England noted that Boston had three Baptist churches, two Methodist, two Episcopalian, one Catholic, one Universalist, and one Quaker meetinghouse, in addition to a dozen Congregational churches.[5] By the following year, "dissenting" churches in Connecticut outnumbered the established ones.[6] Their rapid rise coincided with the triumph of Republicanism in the Nutmeg State.[7] This diversification of religious life was in line with the freedom and egalitarianism espoused during the Revolution.[8] In New England, it spelled the end of Puritan hegemony. In

---

[1] Monroe was the first president to visit this district of Massachusetts, soon to become a state. See "The President," *New Bedford Mercury*, 25 July 1817.

[2] George ___, "Letter II: From a Graduate to His Father," *Times* (Hartford), 12 August 1817.

[3] George Mason, *The Life of James Monroe* (Boston: Small, Maynard and Co. 1921), 367-9.

[4] Untitled article, *Times* (Hartford), 8 July 1817.

[5] Henry B. Fearon, *Sketches of America: A Narrative of a Journey of Five Thousand Miles through the Eastern and Western States* (London: Longman, Hurst, Rees, Orme, and Brown, 1818), 115.

[6] William G. McLoughlin, *New England Dissent, 1630-1833*, vol. 2, *The Baptists and the Separation of Church and State* (Cambridge: Harvard University Press, 1971), 919.

[7] Religious minorities had built an alliance with their political counterpart, the "Tolerationists." See, for example, "To the Friends of Toleration," *Connecticut Herald*, 27 May 1817. The anonymous author of this piece claimed all of these groups had been "doomed to a common humiliation" by the Federalist/Congregationalist monopoly on power.

[8] Nathan O. Hatch, *The Democratization of American Christianity* (New Haven:

quick order, several states — New Hampshire in 1816, Connecticut in 1818, Maine and New Hampshire in 1819, (but not Massachusetts until 1833) — voted to "disestablish" the Congregationalist Church and thereby bring their constitutions into line with the First Amendment.[1] When Timothy Dwight, Connecticut's Federalist-Congregationalist "pope," died on January 11, 1817, a centuries-old ecclesiastical dynasty went to its grave as well.[2] As has been discussed above, a factor contributing to its downfall was the failure of this established church adequately to explain or respond to the crop failures of 1816 and preceding years. This shortcoming undermined its authority. However, while Congregationalist ministers were losing popularity, "dissenting" Protestant preachers who claimed they could interpret the bad weather as a sign of divine unhappiness attracted large crowds.[3]

For, despite all these concerns about their relationship with the Almighty, New Englanders were not prepared to turn their backs on Him. Following the "year without a summer" there was no rush to atheism. In England, Shelley and his circle remained on the radical fringe. Few Americans felt — let alone professed — that God did not exist. 1816 came to be generally regarded as a puzzling aberration from an otherwise reliable seasonal pattern. Faith in God as a protecting father overseeing the fate of humanity and Nature was restored — at least temporarily.

Indeed, the years after 1816 saw a new wave of emotionally-charged revivalism. In part, this was a reaction to Deism: many felt that Enlightenment rationality was supplanting traditional Christianity.[4] Enthusiasm for a populist form of religion, which emphasized personal piety and salvation over biblical exegesis or clerical guidance, engulfed the country. Particularly on the frontier, where democracy could more easily take root, this Second Great Awakening won many fervent adherents. Because they saw their grassroots movement as part of God's plan, these new believers were disinclined to look for "natural" explanations for phenomena like the

---

Yale University Press, 1989), 5.

[1] For legal details of this change in religious law, see Carl H. Esbeck, "Dissent and Disestablishment: The Church-State Settlement in the Early American Republic," *Brigham Young University Law Review* 2004:4 (2004): 1524-38. http://www.law2.byu.edu/lawreview4/archives/2004/4/6ESB-FIN.pdf.

[2] In 1800, all of Connecticut churches were Congregationalist. Half a century later, only twenty nine percent were. Reverend Henry Jones, "On the Rise, Growth and Comparative Relations of Other Evangelical Denominations in Connecticut to Congregationalism," *Contributions to the Ecclesiastical History of Connecticut*, vol. 1 (New Haven: William L. Kingsley, 1861), 269.

[3] Stewart H. Holbrook, *The Yankee Exodus: An Account of Migration from New England* (New York: Macmillan, 1950), 77.

[4] Barry Hankins, *The Second Great Awakening and the Transcendentalists* (Westport CT: Greenwood, 2004), 3-4.

weather.[1] For example, when a lightning bolt struck the cupola of the First Presbyterian Church in Lexington, Kentucky, in July 1817, at the start of the Wednesday evening service, and electricity shot down to the ground floor of the meeting house, killing two ladies of exemplary moral character, the rest of the congregation bowed their heads in homage to the power of Almighty God, "who holds the lightnings [sic] in his hand, and directs them where to strike." The fact that another woman in the congregation, who had been deaf in one ear for thirty years, suddenly regained her hearing seemed only to confirm that this event was divinely instigated. Admonished a writer for the *Western Monitor:* "When all those circumstances are taken in connexion [sic] with the melancholly [sic] event and duly considered — can any one [sic] be so skeptical, so faithless, as not to believe that the whole series of causes and events, in the natural and moral world, are constantly under the directing and controlling Providence of an all-wise God." One churchgoer, who had witnessed the electrocution, told her husband that she now wished to make a declaration of her newfound faith.[2]

In New England, preachers like Lyman Beecher, alarmed by the rise of "Toleration" thinking, decided to go on the offensive against Republican-backed "infidelity."[3] Speaking at the Boston ordination of the fifth son of his recently deceased Yale mentor, Timothy Dwight, in September 1817, Beecher bluntly confronted the Deist forces then gaining momentum in his state. God's laws, he affirmed, could not be fathomed by reason because they did not operate according to reason. Thus it was futile for anyone to read the Bible with hopes of understanding why the Deity acted the way He did: "The mariner who can rectify his disordered compass by his intuitive knowledge for the polar direction need not first rectify his compass, and then obey its direction; he may throw it overboard, and without a luminary of heaven, amid storms and waves and darkness, may plough the ocean, guided only by the light within."[4] In this atmosphere, those who spoke about "natural" causes or sought to study the weather scientifically were put on the defensive.[5]

---

[1] However, this revivalism was not completely hostile to the methods of science. The young Presbyterian preacher Barton Stone, for example, used Bacon's method of objective inquiry to study and write about the emotional effects of attending camp meetings in Kentucky. Hankins, *Awakening*, 10-11.

[2] "General Miscellany," *Dedham Gazette* (Dedham MA), 5 September 1817. Reprinted from the *Western* (Kentucky) *Monitor*.

[3] See Beecher's April 1817 letter to Nathaniel Taylor, quoted in Charles R. Keller, *The Second Great Awakening in Connecticut* New York: Archon, 1968), 60-1.

[4] Beecher, Sermon V, "The Bible a Code of Laws," 3 September 1817, *Beecher's Works*, vol. 2, 171.

[5] 29 Although he disavowed the divine revelations of Christianity on board the *Beagle* in the late 1830s, Charles Darwin held off publishing his theory about the origin of species for more than two decades, largely out of consideration for the religiosity of his wife, Emma.

Protestations that storms, rain, drought, and earthquakes could not possibly occur without some higher meaning or purpose lasted through the 19th century. A natural philosopher like Bowdoin's Parker Cleaveland remained "half disposed" to believe that it rained heavily for several years in a row on the college's new commencement day in early August because this was a *dies infaustus* (unlucky day) — vindication of his stubborn refusal to change the date.[1] As long as he was still able, Chester Dewey took delight in roaming the fields and mountains near his home, rock hammer in hand, specimen bag slung over his shoulder, convinced that by so investigating Nature one could best study God — in "that book of Revelation whose leaves are the fields, and caroling birds the commentators."[2] When a prolonged drought and cold spell in the Ohio Valley and New England finally abated in the late 1830s, The Reverend Hubbard Winslow, Lyman Beecher's successor in the pulpit of Boston's Bowdoin Street Church, praised a "kind Providence" for finally allowing wheat to ripen and granaries to fill. Winslow asked his parishioners to join him in thanking God for his "benevolent operations" and in acknowledging their sins in return for the food now granted them.[3]

Just as fundamental tenets of Christianity experienced a renaissance, so did older values and attitudes in other spheres of life. In the arts, the solitary intensity of Romantic egoism — personified by tormented heroic figures such as Shelley's Prometheus — gave way to a more benign, unifying view of Nature. Ralph Waldo Emerson built Transcendentalism on the notion of a "common soul" connecting all human beings to an "all-loving" natural world.[4] All objects dwelling within it "make a kindred impression," he wrote in his seminal 1836 essay "Nature," as long as the "mind is open to their influence." Henry David Thoreau sensed a similarly uplifting affinity for the woods surrounding Walden Pond, where, during his 1845–1846 solitary sojourn, he reflected that, cut off from Nature, "men remain in their present low and primitive condition; but if they should feel the influence of the spring of springs arousing them, they would of necessity rise to a higher and more ethereal life."[5] The flowering of the Transcendental movement in New England during the 1840s owed much to the early lyrical poems of

---

[1] Leonard Woods, *Address on the Life and Character of Parker Cleaveland, LL.D.* 2nd ed. (Brunswick ME: Joseph Griffin, 1860), 64.

[2] Henry Fowler, *The American Pulpit: Sketches, Biographical and Descriptive, of Living American Preachers* (New York: Fairchild, 1856), 63.

[3] Hubbard Winslow, *Rejoice with Trembling: A Discourse Delivered in Bowdoin Street Church, Boston, on the Day of Annual Thanksgiving, November 30, 1837*, 2nd ed. (Boston: Perkins and Marvin, 1838), 12, 14.

[4] This phrase comes from Emerson's 1847 poem "The World-Soul." Emerson's notion of unity borrows from Hegel's 1807 *The Phenomenology of Spirit*, in which he describes a *Weltgeist*, or "spirit of the world."

[5] Henry D. Thoreau, *Walden* (Boston: Shambhala, 2004), 22.

Wordsworth, written over three decades before, and created a way back from the existential abyss glimpsed by Shelley and his disciples. A parallel converging of human consciousness and physical world would manifest itself in the visual arts — in the landscapes of Hudson River School artists such as Thomas Cole, Jasper Cropsey, Frederic Church, and Albert Bierstadt.

In the economic arena, New England's conversion to a manufacturing center accelerated after 1816, as political and business leaders saw this transformative process as essential to increasing the region's standard of living and attractiveness as a place to live.[1] By 1840, nearly fifteen percent of the labor force — some one hundred thousand persons — was employed in factories or workshops, up from only about one percent in 1816.[2] On the eve of the Civil War, that number would stand at 386,000 — 260,000 men and 126,000 women — or more than a third of all persons who held jobs. Meanwhile, the percentage of New Englanders making a living as farmers kept declining. Samuel Slater's dream of making the United States the equal of England in manufacturing was realized. So were the hopes of New Englanders that the exodus westward would slow or stop.

These various — and often incongruous — developments after 1816 illustrate some guiding assumptions of this book: first of all, history does not "progress" in an orderly, straightforward manner, like rows of grenadiers marching unwaveringly across a parade ground; secondly, single events, no matter how momentous, only rarely alter the course of human affairs, and generally only accelerate it, as the pull of a planet's gravity can speed up a passing comet; and, thirdly, the concept of "cause and effect," central to the quest for knowledge in the natural sciences, is largely not germane to the search for historical "truth" since past events are unique, arising from a myriad of circumstances, the sum total of which we can never fully comprehend. For all of these qualifying reasons, it cannot be demonstrated that the "year without a summer" was solely accountable for the major shifts in religious belief, scientific thinking, economic organization, political ideology, philosophical outlook, aesthetic theory, and demographics outlined in these pages. At most, this abrupt, alarming change in climate gave momentum to trends which had been developing for some time, especially in New England. These include the questioning of hierarchical authority (in church and state); the spread of democracy and dissent; the growing interest in observing *and* theorizing about natural phenomena; the rise of an individualistic ethos in thought and expression; the advent of the Industrial Age; and the shift of America's center from the Eastern seaboard to the Midwestern heartland.

---

[1] See, for example, "Emigration," *Columbian* (New York), 13 October 1817, in *Peopling of New Connecticut*, 22.

[2] Robert B. Zevin, *The Growth of Manufacturing in Early Nineteenth Century New England* (New York: Arno Press, 1975), 10-11.

In their attempts to understand the natural world and the causes of earthly events, many people still relied on faith to instruct them. Despite important insights into the physical principles governing the movement of winds and other meteorological phenomena, the devout stuck to their conviction that the "ultimate causes of the weather arise from the spiritual states under Divine control," as one late 19ᵗʰ-century English clergyman put it.[1] However, discoveries of the laws of thermodynamics; of the correlation between changes in barometric pressure, wind direction, and storms; about the formation of cyclones and clouds; and concerning causes of atmospheric movement provided a more accurate method of making forecasts. These scientific explanations undercut theological teachings and encouraged the study of meteorology in lieu of prayer.[2] The last obstacle in the path of scientific progress — a dearth of precise, replicable, and revealing information — was removed. By the end of the 19ᵗʰ century, Cardinal Newman's assertion that "there is at present no real science of the weather because you cannot get hold of the facts and truths on which it depends" no longer held true.[3]

Although causal links between the "year without a summer" and these long-term changes may not be readily apparent, there is ample evidence that this highly unusual and damaging weather did embolden new ways of thinking. By the end of the nineteenth century, the critical reassessment of old truths facilitated by new information about the physical world had reshaped the political, economic, religious, and geographical landscape of this country. Tensions between faith and reason, tradition and innovation, authority and individuals, necessity and freedom, countryside and city, East and West, persisted, but the forces of modernism had gained a distinct advantage. In our day, these conflicts continue, as they will no doubt until the end of human civilization, but their contours are decidedly different, in part because of what became vitally important in the northern United States and Western Europe two centuries ago — the need to explain why the world had changed. Following that jarring threat to human well-being, an unquestioning belief in a fixed, unalterable order, created and controlled by a benevolent intelligence, was harder to justify. Instead, the imperative to investigate our world grew stronger. For on the results of this ongoing inquiry hangs our survival on this planet. Knowledge has to be instructive as

---

[1] Reverend Thomas Mackereth, "The Causes of the Weather," *Morning Light* 6:292 (24 August 1883): 304.

[2] By the turn of the century, meteorology was still not widely – or well – taught at the university level in the United States. See Robert DeCourcy Ward, "Current Notes on Meteorology," *Science* 20:512 (21 October 1904): 540.

[3] John Henry, Cardinal Newman, "A Form of Infidelity of the Day," quoted in Wilfrid Ward, *The Life of John Henry, Cardinal Newman*, Vol. 1 (London: Longmans, Green and Co., 1921), 393.

well as prescriptive. Accepting that this need for information is as important for our existence as moral dictates, we have come to see that religion and science have complementary value in our lives. The climate crisis of 1816 taught the importance of integrating the physical and the metaphysical in constructing reality and in responding to its demands. Two hundred years later, as we approach another such crisis, we are still figuring out how to do this.

ACKNOWLEDGEMENTS

Many archivists and librarians across New England and beyond have generously assisted me in gathering materials relating to the "year without a summer." I would like to thank the following: the staff of the Helga J. Ingraham Memorial Library, Litchfield Historical Society, Litchfield CT; Diana McCain, Florence S. Marcy Crofut Head of the Research Center, Connecticut Historical Society, Hartford; staff of the History and Genealogy Unit, Connecticut State Library, Hartford; Debby Shapiro, Executive Director, Middlesex County Historical Society, Middletown CT; Barbara Goodwin, Librarian, Windsor Historical Society, Windsor CT; the reference librarians at the Russell Library, Middletown CT; town historian Katherine Chilcoat and staff of the Scoville Memorial Library, Salisbury CT; Rachel Quish, Collections Manager, Wethersfield Historical Society; reference librarians at the Massachusetts Historical Society, Boston; Jessica Gill, Archivist, Newburyport Archival Center, Newburyport Public Library, Newburyport MA; Barbara Krieger, Archivist, Rauner Special Collections Library, Dartmouth College, Hanover NH; Lee Teverow, Reference Librarian, Rhode Island Historical Society Library, Providence; Jamie Kingman Rice, Public Services Librarian, Maine Historical Society, Portland; Richard Lindeman and his staff at the George J. Mitchell Department of Special Collections and Archives, Bowdoin College Library, Brunswick ME; Beth Carroll-Horrocks, Director of Archives, the American Academy of Arts and Sciences, Philadelphia; and the reference staff at Butler Library, Columbia University, New York.

At Algora Publishing, I am much indebted to Andrea Decker for having shepherded this manuscript into book form so adroitly.

In addition to these professionals, I would like to extend a special thanks to my friend and summer neighbor in Brooklin, Maine — Terry Mason — who first told me about the remarkable weather which New England had endured in 1816 and whetted my appetite to learn more.

Salisbury, Connecticut

BIBLIOGRAPHY

Adams, Eliphalet. *God Sometimes Answers His People, by Terrible Things in Righteousness.* London: T. Green, 1735.

Adams, George. *A Short Dissertation on the Barometer, Thermometer, and Other Meteorological Instruments.* London: R. Hindmarsh, 1790.

Adams, John, and Thomas Jefferson. *The Adams-Jefferson Letters.* Edited by Lester J. Cappon. New York NY: Simon & Schuster, 1959.

Adams, Sherman W. *The History of Ancient Wethersfield,* Vol. 1. New York: Grafton, 1904.

Ahrens, C. Donald. *Meteorology Today: An Introduction to Weather, Climate, and the Environment.* 8th ed. Belmont CA: Thompson, 2007.

Allen's New-England Almanack, for the Year of Our Lord 1816. Calculated for the Horizon and Meridian of Hartford, Lat. 41 Deg 50 Min. North. Hartford: Peter Gleason, 1815.

An Agricultural and Economical Almanack, for the Year of Our Lord 1816. Calculated for the Meridian of New Haven — Lat. 41, 18.' New Haven: Society for Promoting Agriculture in the State of Connecticut, 1815.

Anderson, Virginia. *New England's Generation: The Great Migration and the Formation of Society and Culture in the Seventeenth Century.* Cambridge: Cambridge University Press, 1991.

Appleton, Jesse. *Addresses by Rev. Jesse Appleton, D.D., with a Sketch of His Character.* Brunswick ME: Joseph Griffin, 1820.

_____. *Lectures Delivered at Bowdoin College and Occasional Sermons.* Brunswick ME: Joseph Griffin, 1822.

Arnold, Seth Shaler. "'As the Years Pass' — The Diaries of Seth Shaler Arnold

(1788-1871), A Vermonter." *Vermont History* 8, no. 2 (June 1940): 107-93.

Arthur, Brian. *How Britain Won the War of 1812: The Royal Navy's Blockades of The United States, 1812-1815.* Woodbridge, Suffolk: Boydell Press, 2011.

Atack, Jeremy, and Fred Bateman. *To Their Own Soil: Agriculture in the Antebellum North.* Ames: Iowa State University Press, 1987.

Austin, George Lowell. *The History of Massachusetts, from the Landing of the Pilgrims to the Present Time.* Boston: B.B. Russell, 1876.

Bailey, James M. *History of Danbury, Conn., 1684-1896.* New York: Burr, 1896.

Banks, Ronald F. "The Maine Constitutional Convention of 1819." In *A History of Maine: A Collection of Readings on the History of Maine, 1600-1976,* edited by Ronald F. Banks, 4th ed., 179-94. Dubuque IA: Kendall/Hunt, 1976.

_____. "The September Election and the Brunswick Convention of 1816." In *A History of Maine: A Collection of Readings on the History of Maine, 1600-1976,* edited by Ronald F. Banks, 4th ed., 166-78. Dubuque: Kendall/Hunt, 1976.

_____. "The War of 1812: A Turning Point in the Movement to Separate Maine from Massachusetts." In *A History of Maine: A Collection of Readings on the History of Maine, 1600-1976,* edited by Ronald F. Banks, 4th ed., 161-65. Dubuque: Kendall/Hunt, 1976.

Banta, Richard E. *The Ohio.* New York: Henry Holt, 1949.

Baron, W.R. "Historical Climate Records from the Northeastern United States, 1640 to 1900." In *Climate Since 1500,* edited by Raymond S. Bradley and Philip D. Jones, 74-91. London, New York: Routledge, 1992.

Baron, William R. "1816 in Perspective: The View from the Northeastern United States." In *The Year Without a Summer? World Climate in 1816,* edited by C.R. Harington, 124-44. Ottawa: Canadian Museum of Nature, 1992.

Barron, Hal S. *Those Who Stayed Behind: Rural Society in Nineteenth-Century New England.* Interdisciplinary Perspectives on Modern History. Edited by Robert Fogel and Stephan Thernstrom. Cambridge: Cambridge University Press, 1984.

Barstow, George. *History of New Hampshire, From Its Discovery, in 1614, to the Passage of the Toleration Act, in 1819.* Concord: I.S. Boyd, 1842.

Bassett, T.D. Seymour. "The Rise of Cornelius Peter Van Ness, 1782-1826." *Vermont History* 10, no. 1 (March 1942): 3-20.

Bate, Jonathan. *The Song of the Earth.* Cambridge: Harvard University Press, 2000.

Battle, J.H. *History of Morrow County and Ohio: Containing a Brief History of the State of Ohio.* Chicago: O.L. Baskin, 1880.

Beecher, Lyman. *Autobiography, Correspondence of Lyman Beecher, D.D.* Vol. 1.

Edited by Charles Beecher. New York: Harper & Brothers, 1864.

_____. *Beecher's Works.* Vol. 2. *Sermons Delivered on Various Occasions.* Boston: John P. Jewett, 1852.

Bell, Michael B. "Did New England Go Downhill?" *Geographical Review* 79, no. 4 (October 1989): 450-66.

Belville, John Henry. *A Manual of the Thermometer, Containing Its History and Use as a Meteorological Instrument.* London: Richard and John Edward Taylor, 1850.

Ben-David, Joseph. *American Higher Education: Directions Old and New.* Berkeley: Carnegie Foundation for the Advancement of Teaching, 1972.

Bentley, William. *The Diary of William Bentley, Pastor of the East Church, Salem, Massachusetts.* Vol. 4. Salem: Essex Institute, 1914.

Berryman, Charles. *From Wilderness to Wasteland: The Trial of the Puritan God in the American Imagination.* Port Washington NY: Kennikat Press, 1979.

Bidwell, Percy W. "Rural Economy in New England at the Beginning of the Nineteenth Century." *Transactions of the Connecticut Academy of Arts and Sciences* 20 (April 1916): 241-339.

_____."The Agricultural Revolution in New England." *American Historical Review* 26, no. 4 (July 1921): 683-702.

Blodgett, Lorin. *Climatology of the United States.* Philadelphia: J.B. Lippincott, 1857.

Boney, F.N. *A Pictorial History of the University of Georgia.* Athens: University of Georgia Press, 1984.

Bowditch, Nathaniel. *Memoir of Nathaniel Bowditch.* Boston: James Munroe, 1841.

_____. "Observations of the Comet of 1807." *Memoirs of the American Academy of Arts and Sciences* 3, no. 1 (1809): 1-17.

Bremer, Francis J. *John Winthrop: America's Forgotten Founding Father.* New York: Oxford University Press, 2003.

Broadcloth, Jonathan [pseud.]. *The United States' Political Looking-Glass, Hydrometer and Thermometer.* Albany NY: [publisher not identified], 1824.

Buchan, Alexander. *Introductory Text-Book of Meteorology.* Edinburgh: Blackwood, 1871.

Burke, Edmund. *A Philosophical Enquiry into the Origin of Our Ideas of the Sublime and the Beautiful.* 5th ed. London: J. Dodsley, 1767.

Burt, Christopher C. *Extreme Weather: A Guide and Record Book.* New York: Norton, 2004.

Bushman, Richard Lyman. *From Puritan to Yankee: Character and the Social Order* in Connecticut, 1690-1765. Cambridge: Harvard University Press, 1967.

_____. *Joseph Smith: Rough Stone Rolling. A Cultural Biography of Mormonism's Founder*. New York: Knopf, 2005.

Byron, Lord. *Poetical Works.* Edited by Frederick Page. 3rd ed. London: Oxford University Press, 1973.

Campbell, Charlie. *Scapegoat: A History of Blaming Other People*. New York: Duckworth Overlook, 2011.

Carter, Horace S. "Josiah Meigs, Pioneer Weatherman." *Weatherwise* 13 (1960): 166-67.

Catchpole, A.J.W. "River Ice and Sea Ice in the Hudson Bay Region during the Second Decade of the Nineteenth Century." In *The Year Without a Summer? World Climate in 1816*, edited by C.R. Harington, 233-44. Ottawa: Canadian Museum of Nature, 1992.

Chalmers, Thomas. *A Series of Discourses on the Christian Revelation*. Appendix No. 1: "Astronomical Discourses." Glasgow: Collins, 1818.

Channing, William Ellery. *Memoir of William Ellery Channing*. Vol. 1. 3rd ed . London: John Chapman, 1848.

_____."Extracts from Sermons Preached on Days of Humiliation and Prayer Appointed in Consequence of the Declaration of War Against Great Britain, A.D. 1812." In Channing, *Discourses, Reviews, and Miscellanies*, 583-90. Boston: Gray and Bowen, 1830.

_____.*The Works of William E. Channing, D.D.* Vol. 3. 11th ed . Boston: Walker, Fuller, 1866.

Channing, William Henry. *The Life of William Ellery Channing, D.D.* Boston: American Unitarian Association, 1880.

Chase, Benjamin. *History of Old Chester (N.H.) from 1719 to 1869*. Auburn NH: Benjamin Chase, 1869.

Chauncy, Charles. *The Mystery Hid from Ages and Generations Made Manifest by the Gospel-Revelation*. London: Charles Dilly, 1784.

Clairmont, Claire. *The Clairmont Correspondence: Letters of Claire Clairmont, Charles Clairmont, and Fanny Imlay Clairmont*, Vol. 1, *1808-1834*. Edited by Marion Kingston Stocking. Baltimore: Johns Hopkins University Press, 1995.

Clark, Christopher. *The Roots of Rural Capitalism: Western Massachusetts, 1780-1860*. Ithaca: Cornell University Press, 1990.

Clark, George L. *History of Connecticut: Its People and Intuitions*. New York: G.P. Putnam's Sons, 1914.

Clark, Victor S. *History of Manufactures in the United States, 1607-1860*. Washington: Carnegie Institution of Washington, 1916.

Cleaveland, Parker. "Meteorological Observations, Made at Bowdoin College." *Memoirs of the American Academy of Arts and Sciences* 3, no. 1 (1809): 119-21.

Clerc, Laurent. *Diary of Laurent Clerc's Voyage from France to America in 1816.* Hartford: American School for the Deaf, 1952.

Clubbe, John. "Tempest-Toss'd Summer of 1816: Shelley's *Frankenstein.*" *Byron Journal* 91 (1991): 26-40.

Cogliano, Francis D. *Thomas Jefferson: Reputation and Legacy.* Charlottesville: University of Virginia Press, 2006.

Cohen, I. Bernard. *Science and the Founding Fathers: Science in the Political Thought of Thomas Jefferson, Benjamin Franklin, John Adams and James Madison.* New York: Norton, 1995.

Cowley, W.H., and Don Williams. *International and Historical Roots of American Higher Education.* New York: Garland, 1991.

Cox, John D. *Storm Watchers: The Turbulent History of Weather Prediction from Franklin's Kite to El Nino.* Hoboken NJ: Wiley, 2002.

Crandall, Ralph J. "New England's Migration Fever: The Expansion of America." *Ancestry* (July-August 2000): 15-19.

Croce, Benedetto. *History of Europe in the Nineteenth Century.* Translated by Henry Furst. New York: Harcourt, Brace & World, 1963.

Crockett, Walter Hill. *Vermont: Green Mountain State.* Vol. 3. New York: Century History, 1921.

Crombie, Alexander. *Natural Theology, or Essays on the Existence of Deity and of Providence.* Vol. 1. London: G. Woodfall, 1829.

Dalton, John. *Meteorological Observations and Essays.* 2nd ed. London: Baldwin and Cradock, 1834.

Danhof, Clarence H. *Change in Agriculture: The Northern United States, 1820-1870.* Cambridge: Harvard University Press, 1969.

Davison, Charles. *A History of British Earthquakes.* Cambridge: Cambridge University Press, 2009.

Dewey, Chester. "Result of Meteorological Observations, Made at Williams College." *North American Review* 5, no. 13 (21 May 1817): 152-6.

Dick, Thomas. "On Natural Theology." Document III. 8. *Methodist Magazine,* April 1825.

Dike, Nathaniel. "Nine Letters of Nathaniel Dike on the Western Country, 1816-1818." *Ohio History* 67 (July 1958): 189-220.

Donahue, Brian. *The Great Meadow: Farmers and the Land in Colonial Concord.* New Haven: Yale University Press, 2004.

Dray, Philip. *Stealing God's Thunder: Benjamin Franklin's Lightning Rod and the Invention of America.* New York: Random House, 2005.

Dupree, A. Hunter. *Science in the Federal Government.* Cambridge MA: Belknap Press, 1957.

Dwight, Henry E. *Travels in the North of Germany: In the Years 1825-1826.* New York: G. & C. & H. Carvill, 1829.

Dwight, Timothy. *A Statistical Account of the City of New-Haven.* New Haven: Walter and Steel, 1811.

_____.*Theology Explained and Defended.* Vol. 1. Bedford MA: Applewood Books, 1848.

_____.*Travels in New England and New York.* Vol. 4. New Haven: T. Dwight, 1822.

Eather, Robert H. *Majestic Lights: The Aurora in Science, History, and the Arts.* Washington: American Geophysical Union, 1980.

Eddy, John A. "Before Tambora: The Sun and Climate, 1790-1830." In *The Year Without a Summer? World Climate in 1816,* edited by C.R. Harington, 11-12. Ottawa: Canadian Museum of Nature, 1992.

Elliott, Amanda. Diary of Amanda Elliott, 920 EL 425. 1816. File: Typescript: [Photocopy]. History and Genealogy, Connecticut State Library. Hartford.

Elliott, Orrin L. *The Tariff Controversy in the United States, 1789-1833.* Palo Alto: Stanford University, 1892.

Emerson, Ralph. *A Sermon Preached at Norfolk, Connecticut, May 16, 1816: The First Sabbath after His Ordination.* Hartford: George Goodwin & Sons, 1817.

Emery, Caleb. "The Cold Summer of 1816." In *Collections, Historical and Miscellaneous, and Monthly Literary Journal,* 254-62. Vol. 2. Edited by J. Farmer and J. B. Moore, 254-302. Concord NH: J.B. Moore, 1823.

Emmons, Nathanael. *Sermons on Various Subjects of Christian Doctrine and Duty.* Vol. 5. Providence: Barnum Field, 1825.

_____.*The Works of Nathanael Emmons, D.C., with a Memoir of His Life.* Vol. 3. Edited by Jacob Ide. Boston: Crocker & Brewster, 1842.

Fagan, Brian. *The Little Ice Age: How Climate Made History, 1300-1850.* New York: Basic Books, 2000.

Fearon, Henry Bradshaw. *Sketches of America: A Narrative of a Journey of Five Thousand Miles Through the Eastern and Western States.* London: Longman, Hurst, Rees, Orme, and Brown, 1818.

Field, David D. *Statistical Account of the County of Middlesex in Connecticut.* Middletown: Connecticut Academy of Arts & Sciences, 1819.

Field, David D., and Chester Dewey. *A History of the County of Berkshire,*

Massachusetts, in Two Parts. Pittsfield: S. W. Bush, 1829.

Fishlow, Albert. *American Railroads and the Transformation of the Ante-Bellum Economy*. Cambridge: Harvard University Press, 1965.

Fleming, James R. *Historical Perspectives on Climate Change*. New York: Oxford University Press, 1998.

Flint, Charles L. *Fourteenth Annual Report of the Secretary of the Massachusetts Board of Agriculture*. Boston: Wright & Potter, 1867.

Fowler, Henry. *The American Pulpit: Sketches, Biographical and Descriptive, of Living American Preachers, and of the Religious Movements and Distinctive Ideas Which They Represent*. New York: J. M. Fairchild, 1856.

Francis, W. H. *History of the Hatting Trade in Danbury, Conn. From Its Commencement in 1780 to the Present Time*. Danbury: H. & L. Osborne, 1860.

Franklin, Benjamin. *The Autobiography of Benjamin Franklin*. New York: Henry Holt, 1916.

_____.*Experiments and Observations on Electricity Made at Philadelphia in America, by Benjamin Franklin*. 4th ed. London: D. Henry, 1769.

Friedman, John S. *Out of the Blue: A History of Lightning: Science, Superstition, and Amazing Stories of Survival*. New York: Random House, 2008.

Fuchs, Karl. "The Great Earthquakes of Lisbon 1755 and Aceh 2004 Shook the World: Seismologists' Societal Responsibility." In *The 1755 Lisbon Earthquake: Revisited*, edited by Luis A. Mendez-Victor, Carlos Sousa Oliveira, Joao Azevedo and Antonio Ribeiro, 43-64. New York: Springer, 2005.

Fuller, Grace P. "An Introduction to the History of Connecticut as a Manufacturing State." *Smith College Studies in History* 1, no. 1.(October 1915): 1-64.

Geiger, Roger L. "New Themes in the History of Nineteenth-Century Colleges." In *The American College in the Nineteenth Century*, edited by Roger L. Geiger, 1-36. Nashville: Vanderbilt University Press, 2000.

Gold, Thomas. *Address of Thomas Gold, Esq., President of the Berkshire Agricultural Society, Delivered Before the Berkshire Association for the Promotion of Agriculture and Manufactures, at Pittsfield, Oct. 2d, 1817*. Pittsfield: Phinehas Allen, 1817.

_____.*Address of Thomas Gold, Esq., President of the Berkshire Agricultural Society and Member of the Massachusetts Society, Delivered Before the Berkshire Association for the Promotion of Agriculture and Manufactures, at Pittsfield, Oct. 3d, 1816*. Pittsfield: Phinehas Allen, 1816.

Goode, George Brown. *The Origins of the National Scientific and Educational Institutions of the United States*. New York: G.P. Putnam's Sons, 1890.

Goodrich, Carter, et al. *Canals and American Economic Development*. Port Washington NY: Kennikat Press, 1961.

Goodrich, Samuel G. *Recollections of a Life Time*. Vol. 2. New York: Miller, Orton & Mulligan, 1856.

Grove, Jean M. *The Little Ice Age*. London: Routledge, 1990.

Handlin, Oscar, and Mary F. Handlin. *The American College and American Culture: Socialization as a Function of Higher Education*. New York: McGraw-Hill, 1970.

Hascall, Daniel. *Caution Against False Philosophy: A Sermon*. Hamilton NY: Johnson & Sons, 1817.

Hatch, Nathan O. *The Democratization of American Christianity*. New Haven: Yale University Press, 1989.

Hatcher, Harlan. *The Western Reserve: The Story of New Connecticut in Ohio*. Indianapolis: Bobbs-Merrill, 1949.

Hawley, Stephen. *The Agency of God in Snow, Frost, Etc. A Discourse, Delivered at Bethany, in New-Haven, January 6th, 1771*. New Haven: Thomas & Samuel Green, 1771.

Heidler, David S., and Jeanne T. Heidler. *Daily Life in the Early American Republic, 1790-1820: Creating a New Nation*. Westport: Greenwood Press, 2004.

Heidorn, Keith C. "Eighteen Hundred and Froze to Death: The Year There Was No Summer." *Weather Doctor*. 1 July 2000. Web. 22 August 2010.

History of Concord, New Hampshire, from the Original Grant in Seventeen Hundred and Twenty-Five to the Opening of the Twentieth Century. Edited by James O. Lyford. Concord: Rumford Press, 1903.

History of Connecticut in Monographic Form. Edited by Norris G. Osborn. New York: States History Co., 1925.

History of Middlesex County, Connecticut, with Biographical Sketches of Its Prominent Men. New York: J.B. Beers, 1884.

Hoey, John B. "Federalist Opposition to the War of 1812." *Early American Review* 3, no. 1 (Winter 2000). Web. 16 Sept. 2010.

Hofstadter, Douglas, and C. DeWitt Hardy. *The Development and Scope of Higher Education in the United States*. New York: Columbia University Press, 1952.

Hoke, Donald R. *Ingenious Yankees: The Rise of the American System of Manufactures in the Private Sector*. New York: Columbia University Press, 1990.

Holbrook, Stewart H. *The Yankee Exodus: An Account of Migration from New England*. New York: MacMillan, 1950.

Holland, Josiah G. *History of Western Massachusetts: The Counties of Hampden,*

Hampshire, Franklin, and Berkshire. Springfield: S. Bowles, 1855.

Hollister, Hiel. *Pawlet for One Hundred Years.* Albany: J. Munsell, 1867.

Holmes, Richard. *The Age of Wonder: How the Romantic Generation Discovered the Beauty and Terror of Science.* New York: Pantheon, 2008.

_____. *Shelley: The Pursuit.* New York: New York Review of Books, 1974.

Holyoke, Edward A. "A Proposal for Adjusting a New Scale to the Mercurial Thermometer." *Memoirs of the American Academy of Arts and Sciences* 3, no. 1 (1809): 51-6.

Holyoke, Edward A., and Enoch Hale. "A Meteorological Journal from the Year 1786 to the Year 1829, Inclusive, with a Prefatory Memoir." *Memoirs of the American Academy of Arts and Sciences* 1 (1 January 1833): 107-216.

Hornberger, Theodore. "The Science of Thomas Prince." *New England Quarterly* 9, no. 1 (March 1936): 26-42.

Horowitz, Helen L. *Campus Life: Undergraduate Cultures from the End of the Eighteenth Century to the Present.* New York: Knopf, 1987.

Horton, James Oliver. "Race and Religion: Ohio, America's Middle Ground." In *Ohio and the World, 1753-2054: Essays toward a New History of Ohio,* edited by Geoffrey Parker, Richard Sisson, and William Russell Coil, 43-72. Columbus, Ohio: State University Press, 2005

Horton, Wesley W. "Connecticut Constitutional History, 1776–1988." August 1988. *Connecticut State Library.* Web. 27 July 2010.

Hounshell, David A. *From the American System to Mass Production, 1800-1932: The Development of Manufacturing Technology in the United States.* Baltimore: Johns Hopkins University Press, 1985.

Hoyt, Douglas V., and Kenneth H. Schatten. *The Role of the Sun in Climate Change.* New York: Oxford University Press, 1997.

Hoyt, Joseph B. "The Cold Summer of 1816." *Annals of the Association of American Geographers* 48, no. 2 (June 1958): 118-31.

Hume, David. *Dialogues Concerning Natural Religion.* London: William Blackwood & Sons, 1907.

_____. *An Essay on Miracles.* London: J. Watson, 1840.

Humphreys, David. *A Discourse on the Agriculture of the State of Connecticut, and the Means of Making It More Beneficial to the State.* New Haven: T.G. Woodward, 1816.

Humphrey, Gregory. "1816 Volcanoes Made Summer Feel Like Winter in Eastern United States." *Caffeinated Politics.* 20 April 2010 . Web. 25 August 2010.

Hurt, R. Douglas. *The Ohio Frontier: Crucible of the Old Northwest, 1720-1830.*

Bloomington: Indiana University Press, 1996.

Imholt, Robert J. "Timothy Dwight, Federalist Pope in Connecticut." *New England Quarterly* 73, no. 3 (September 2000): 386.

Isaacson, Walter. *Benjamin Franklin: A Life.* New York: Simon & Schuster, 2003.

Ives, J. Moss. "Connecticut in the Manufacturing World." *Connecticut Magazine* 7, no. 5 (February/March 1902): 627-46.

Jackson, Leon. "The Rites of Man and the Rites of Youth: Fraternity and Riot at Eighteenth-Century Harvard." In *The American College in the Nineteenth Century*, edited by Roger L. Geiger, 46-79, Nashville: Vanderbilt University Press, 2000.

Jankovic, Vladimir. *Reading the Skies: A Cultural History of English Weather, 1650-1820.* Chicago: University of Chicago Press, 2001.

Jefferson, Thomas. *The Works of Thomas Jefferson.* New York: G.P. Putnam's Sons, 1905.

Jerome, Chauncey. *History of the American Clock Business for the Past Sixty Years, and a Life of Chauncey Jerome, Written by Himself.* New Haven: Drayton, 1860.

Johnson, Curtis D. *Islands of Holiness: Rural Religion in Upstate New York, 1790-1860.* Ithaca, London: Cornell University Press, 1989.

Jones, Rev. Henry. "On the Rise, Growth and Comparative Relations of Other Evangelical Denominations in Connecticut to Congregationalism." In *Contributions to the Ecclesiastical History of Connecticut.* Vol. 1, 260-73. New Haven: William L. Kingsley, 1861.

Kafer, Peter K. "The Making of Timothy Dwight: A Connecticut Morality Tale." *William and Mary Quarterly* 47, no. 2 (April 1990): 189-209.

Keller, Charles Roy. *The Second Great Awakening in Connecticut.* New York: Archon, 1968.

Kelley, Brooks Mather. *Yale: A History.* New Haven: Yale University Press, 1999.

Kellogg, Ebenezer. "Sermons of Rev. Ebenezer Kellogg, Pastor of First Congregational Church, Vernon, Conn., 1762-1818." Connecticut Sermons, 252 K291. 1816. History and Genealogy, Connecticut State Library, Hartford

King, Rufus. *Ohio: First Fruits of the Ordinance of 1787.* Boston: Houghton Mifflin, 1888.

Kington, John. "Weather Patterns over Europe in 1816." In *The Year Without a Summer? World Climate in 1816*, edited by C.R. Harington, 358-71. Ottawa: Canadian Museum of Nature, 1992.

Klingaman, William K., and Nicholas P. Klingaman. *The Year Without Summer:*

1816 and the Volcano That Darkened the World and Changed History. New York: St. Martin's Press, 2013.

Klyza, Christopher M., and Stephen C. Trombulak. *The Story of Vermont: A Natural and Cultural History*. Hanover: University Press of New England, 1999.

Knepper, George W. *Ohio and Its People*. Kent: Kent State University Press, 1989.

Kulikoff, Allan. "Migration and Cultural Diffusion in Early America, 1600-1860." *Historical Methods* 19, no. 4 (Fall 1986): 153-69.

Kupperman, Karen O. "The Puzzle of the American Climate in the Early Colonial Period." *American Historical Review* 87, no. 5 (December 1982): 1262-89.

Kuzic, Kresimir. "The Impact of Two Volcano Eruptions on the Croatian Lands at the Beginning of the 19th Century." *Croatian Meteorological Journal* 42 (2007): 15-39.

Lamb, H.H. *Climate, History and the Modern World*. London: Methuen, 1982.

_____.*Climate: Present, Past and Future*. Vol.1. London: Methuen, 1972.

Landsberg, H.E., and J.M. Albert. "The Summer of 1816 and Volcanism." *Weatherwise* 27, no. 2 (April 1974): 63-6.

Lankevich, George J. *River of Dreams: The Hudson Valley in Historic Postcards*. Bronx NY: Fordham University Press, 2006.

Lee, Loyd E. Review of *Hunger in Bayern 1816-1818: Politik und Gesellschaft in einer Staatskrise des frühen 19. Jahrhunderts*, by Gerard Müller. *Central European History* 35, no. 1(March 2002): 118-19 .

Leighton, Angela. *Shelley and the Sublime: An Interpretation of the Major Poems*. Cambridge: Cambridge University Press, 1984.

Levine, George. "The Ambiguous Heritage of Frankenstein." In *The Endurance of Frankenstein: Essays on Mary Shelley's Novel*, edited by George Levine and U.C. Knoepflmacher, 3-30. Berkeley: University of California Press, 1979.

Lilla, Mark. *The Stillborn God: Religion, Politics, and the Modern West*. New York: Knopf, 2007.

Lilly, Alfred T. *The Silk Industry of the United States from 1766 to 1874*. Boston: John Wilson and Son, 1875.

Little, William. *A History of the Town of Warren, New Hampshire, From Its Early Settlement to the Year 1854, Including a Sketch of the Pemigewasset Indians*. Concord: McFarland & Jenks, 1854.

Lodge, Henry Cabot. "The New England States: A Compendious History. The New England States." In *The New England States: Their Constitutional,*

Judicial, Educational, Commercial, Professional and Industrial History, Vol. 1, edited by William T. Davis, 23-7. Boston: D. H. Hurd, 1897.

Lough, J.M. "Climate of 1816 and 1811-20 as Reconstructed from Western North American Tree-Ring Chronologies." In *The Year Without a Summer? World Climate in 1816*, edited by C.R. Harington, 97-114. Ottawa: Canadian Museum of Nature, 1992.

Ludlum, David. *Social Ferment in Vermont, 1791-1850*. New York: AMS Press, 1966.

_____."New England's Dark Day: 19 May 1780." *Weatherwise* 25, no. 3 (June 1972): 112-19.

Madison, James. *The Writings of James Madison*. New York: G.P. Putnam's Sons, 1908.

Mangold, George B. *The Labor Argument in the American Protective Tariff Discussion*. Madison: University of Wisconsin, 1906.

Margo, Robert A. *Wages and Labor Markets in the United States, 1820-1860*. Chicago: University of Chicago Press, 2000.

Marvin, Abijah P. *The Life and Times of Cotton Mather*. Boston: Congregational Sunday-School and Published Society, 1892.

Mather, Cotton. *The Christian Philosopher: Collection of the Best Discoveries in Nature with Religious Improvements*. Charlestown MA: Middlesex, 1815.

McCabe, James D., Jr. *Great Fortunes, and How They Were Made, or The Struggles and Triumphs of Our Self-Made Men*. Cincinnati: Hannaford, 1871.

McCain, Diana Ross. "The Worst Winters." *Early American Life* 25, no. 1 (February 1994): 40-2.

_____."Year Without Summer." *Connecticut*, (July 1987), 48-51.

McClelland, Peter D. *Sowing Modernity: America's First Agricultural Revolution*. Ithaca: Cornell University Press, 1997.

McCoy, Drew R. *The Elusive Republic: Political Economy in Jeffersonian America*. Chapel Hill: University of North Carolina Press, 1980.

McLoughlin, William G. *New England Dissent, 1630-1833*. Vols. 1- 2. *The Baptists and the Separation of Church and State*. Cambridge: Harvard University Press, 1971.

_____. *Rhode Island, A History*. New York: Norton, 1986.

McMahon, Sarah. "A Comfortable Subsistence: The Changing Composition of Diet in Rural New England, 1620-1840." *William and Mary Quarterly* 42, no. 1 (January 1985): 26-65.

_____."Provisions Laid Up for the Family: Toward a History of Diet in New England, 1650-1850." *Historical Methods* 14, no. 1 (Winter 1981): 4-18.

_____."All Things in Their Proper Season': Seasonal Rhythms of Diet in Nineteenth Century New England." *Agricultural History* 63, no. 2 (Spring 1989): 130-51.

Meigs, Josiah. "Response to the Board of Trustees and 'a Candid Community' by Professor Josiah Meigs, 1811." *Hargrett Rare Book & Manuscript Library, University of Georgia Libraries.* 9 August 1811. Web. 15 September 2011.

Meigs, William Montgomery. *Life of Josiah Meigs.* Philadelphia, 1887.

Mellor, Anne K. *Mary Shelley: Her Life, Her Fiction, Her Monsters.* New York: Methuen, 1988.

Melvill, Thomas, Jr. *Address of Thomas Melvill, Jun., Delivered Before the Berkshire Society for the Promotion of Agriculture and Manufactures, at Pittsfield, October 3, 1815.* Pittsfield: Phinehas Allen, 1815.

Melville, Herman. *Moby Dick, or The Whale.* Norwalk CT: Easton, 1977.

Mergen, Bernard. *Snow in America.* Washington DC: Smithsonian Institution Press, 1997.

Meyer, David R. *The Roots of American Industrialization.* Baltimore: Johns Hopkins University Press, 2003.

_____."The Roots of American Industrialization, 1790-1860." *EH.Net.* 1 February 2010. Web. 28 November 2011.

Miles, Henry A. *Lowell, As It Was, and As It Is.* Lowell MA: Powers and Bagley, 1845.

Milham, Willis I. "The Year 1816: The Causes of Abnormality." *Monthly Weather Review* 52, no. 12 (December 1924): 563-70.

Miller, Eric R. "The Evolution of Meteorological Institutions in the United States." *Monthly Weather Review* 59, no. 1 (January 1931).

Milligan, Fred J. *Ohio's Founding Fathers.* Lincoln NE: iUniverse, 2003.

Mishra, Vijay. *The Gothic Sublime.* Albany: State University of New York Press, 1994.

Morison, Samuel Eliot. *The Life and Letters of Harrison Gray Otis, Federalist, 1765-1848.* Vol. 1. Boston: Houghton Mifflin, 1913.

Morley, Maurice. "History Lesson: The Year There Was No Summer in Ballston Spa, Milton, Malta, and Ballston." *Saratogian News.* 15 August 2010. Web. 24 September 2011.

Morris, Charles R. *The Dawn of Innovation: The First American Industrial Revolution.* New York: Public Affairs, 2012.

Morse, Anson Ely. *The Federalist Party in Massachusetts to the Year 1800.* Princeton: University Library, 1909.

Murrin, John M., et al. *Liberty, Equality, Power: A History of the American People.*

Vol. 1. 4th ed. Belmont CA: Thompson/Wadsworth, 2005.

Mussey, Barrows. "Yankee Chills, Ohio Fever." *New England Quarterly* 22, no. 4 (December 1949): 435-51.

Muzzey, David S. *An American History*. Boston, New York: Ginn & Co., 1911.

Nelson Charles A. *Waltham, Past and Present, and Its Industries*. Cambridge MA: Moses King, 1882.

Neumann, Bonnie R. *The Lonely Muse: A Critical Biography of Mary Wollstonecraft Shelley*. Amherst NY: Prometheus Books, 1980.

The New England States: Their Constitutional, Judicial, Educational, Commerical, Professional, and Industrial History. Boston: D.H. Hurd, 1901.

Nolan, Edward J. *A Short History of the Academy of Natural Sciences of Philadelphia*. Philadelphia: Academy of Natural Sciences, 1909.

Norton, Andrews. *Inaugural Discourse, Delivered Before the University in Cambridge, August 10, 1819*. Cambridge MA: Hilliard & Metcalf, 1819.

Ogden, Eliza A. "The Journal of Eliza A. Ogden." Typescript. Litchfield Female Academy Collection, Series 2: Folder 22. 1816-1818. Litchfield Historical Society, Litchfield CT.

Ogilvie, A.E. J. "1816 — Year Without a Summer in Iceland?" In *The Year Without a Summer? World Climate in 1816*, edited by C.R. Harington, 331-54. Ottawa: Canadian Museum of Nature, 1992.

Ogilvie, A.E.J. "Documentary Evidence for Changes in the Climate of Iceland, A.D. 1500 to 1899." In *Climate Since 1500*, edited by Raymond S. Bradley and Philip D. Jones, 92-117. London: Routledge, 1992.

Olmsted, Denison. *An Introduction to Natural Philosophy*. New York: Collins, 1858.

Onuf, Peter S. *Jefferson's Empire: The Language of American Nationhood*. Charlottesville: University of Virginia Press, 2000.

Oppenheimer, Clive. "Climatic, Environmental and Human Consequences of the Largest Known Historic Eruption: Tambora Volcano (Indonesia) 1815." *Progress in Physical Geography* 27, no. 2 (2003): 230-59.

Otis, Harrison Gray. *Otis' Letters in Defense of the Hartford Convention and the People of Massachusetts*. Boston: Simon Gardner, 1824.

Overton, Elizabeth. "The Little Ice Age in New England: A Disaster of Global Proportions or Just Another New England Freeze?" Course paper. 3 May 2007. Vertical File "Connecticut - Weather: Year Without a Summer (1816)," History and Genealogy, Connecticut State Library, Hartford.

O'Brien, Edna. *Byron in Love: A Short Daring Life*. New York: Norton, 2009.

William Paley. *Natural Theology: Or, Evidence of the Existence and Attributes of the*

Deity. London: S. Hamilton, 1813.

Palfrey, John G. *Abstract from the Returns of Agricultural Societies in Massachusetts*. Boston: Dutton and Wentworth, 1846.

Palmer, Benjamin. *Signs of the Times Discerned and Improved*. Charleston: J. Hoff, 1816. Microfiche.

Pant, G.B., B. Parthasarathy, and N.A. Sontakke. "Climate Over India during the First Quarter of the Nineteenth Century." In *The Year Without a Summer? World Climate in 1816*, edited by C.R. Harington, 429-35. Ottawa: Canadian Museum of Nature, 1992.

Parker, Geoffrey. *Global Crisis: War, Climate Change and Catastrophe in the Seventeenth Century*. New Haven: Yale University Press, 2013.

Parton, Ethel. *The Year Without a Summer: A Story of 1816*. New York: Viking, 1945.

Parton, James. "Chauncey Jerome." In *People's Book of Biography, or Short Lives of the Most Interesting Persons of All Ages and Countries*, edited by James Parton, 209-14. New York: Virtue and Yorston, 1868.

Pasley, Jeffrey L. *"Tyranny of Printers": Newspaper Politics in the Early American Republic*. Charlottesville: University of Virginia Press, 2001.

Pease, John, and John M. Niles. *Gazeteer of the States of Connecticut and Rhode-Island, from Original and Authentic Materials*. Hartford: William S. Marsh, 1819.

Perley, Sidney. *Historic Storms of New England*. Salem MA: Salem Press, 1891.

Peskin, Lawrence A. *Manufacturing Revolution: The Intellectual Origins of Early American Industry*. Baltimore: Johns Hopkins University Press, 2003.

Peterson, Edward. *History of Rhode Island*. New York: J.S. Taylor, 1853.

Phillips, Bill. "*Frankenstein* and Mary Shelley's 'Wet Ungenial Summer.'" *Atlantis* 28, no. 2 (December 2006): 59-68.

Pitkin, Timothy. *A Statistical View of the Commerce of the United States: Its Connection with Agriculture and Manufactures*. Hartford: Charles Hosmer, 1816.

Plumer, William. Plumer, William. *An Address to the Clergy of New England, on Their Opposition to the Rulers of the United States*. Concord: I. & W.R. Hill, 1814.

_____. *General Address to the Freemen of New Hampshire, or, The General Government of the New-England Opposition Contrasted*. Concord: [publisher not identified], 1816.

_____. *The Life of William Plumer*. Edited by A.P. Peabody. Boston: Phillips, Sampson, 1857.

_____. *Message from His Excellency the Governor of New Hampshire to the Legislature of New-Hampshire, June Session, 1817*. 5 June 1817. Concord: [publisher not identified], 1817.

Post, John D. *The Last Great Subsistence Crisis in the Western World.* Baltimore: Johns Hopkins University Press, 1977.

_____."The Impact of Climate on Political, Social, and Economic Change: A Comment." *Journal of Interdisciplinary History* 10, no. 4 (Spring 1980): 719-23.

Potts, David B. "Curriculum and Enrollment: Assessing the Popularity of Antebellum Colleges. In *The American College in the Nineteenth Century*, edited by Roger L. Geiger, 37-45. Nashville: Vanderbilt University Press, 2000.

Priestley, Joseph. *The History and Present State of Electricity with Original Experiments.* Vol 1. London: C. Bathurst & T. Lowndes, 1775.

Prince, Thomas. *Earthquakes, the Works of God, and Tokens of His Just Displeasure.* Boston: D. Fowle, 1755.

_____.*Extraordinary Events, the Doings of God, and Marvellous in Pious Eyes;* 3rd ed . Boston: J. Lewis, 1745.

Punter, David, and Glennis Byron. *The Gothic.* Malden MA: Blackwell, 2004.

Purcell, Richard J. *Connecticut in Transition, 1775-1818.* Washington: American Historical Association, 1918.

Purdy, Jedediah. *A Tolerable Anarchy: Rebels, Reactionaries, and the Making of American Freedom.* New York: Knopf, 2009.

Quincy, Josiah. *The History of Harvard University.* Vol. 2. Boston: Crosby, Nicols, Lee, 1860.

Railo, Eino. *Haunted Castle: A Study of the Elements of English Romanticism.* Edinburgh: Edinburgh Press, 1927.

Rampino, Michael R., and Stephen Self. "Historic Eruptions of Tambor (1815), Krakatau (1883), and Agung (1963): Their Stratospheric Aerosols, and Climatic Impact." *Quaternary Research* 18, no. 2 (September 1982): 127-43.

Redfield, William C. *Meteorological Sketches.* New York: E. & G.W. Blunt, 1837.

Reuben, Julie A. *The Making of the Modern University: Intellectual Transformation and the Marginalization of Morality.* Chicago: University of Chicago Press, 1996.

Richard, Henry, and John Carvell Williams. *Disestablishment.* London: Swan Sonnenschein, Le Bas & Lowrey, 1886.

Rieger, James. *The Mutiny Within: The Heresies of Percy Bysshe Shelley.* New York: George Braziller, 1967.

Rind, Bruce, and David Strohmetz. "Effect of Beliefs about Weather Conditions on Tipping." *Journal of Applied Social Psychology* 26, no. 2

(January 1996): 137-47.

Ringwalt, Roland. "Was Webster Inconsistent on the Tariff?" *The Protectionist* 21, no. 249 (January 1910): 483.

Robbins, Thomas. *Diary of Thomas Robbins, D.D.* Vol. 1. *1796-1825.* Edited by Niles Tarbox. Boston: Beacon Press, 1886.

_____. Thomas Robbins Papers, Connecticut Historical Society, Hartford.

Roberts, Jon H., and James Turner. *The Sacred and the Secular University.* Princeton: Princeton University Press, 2000.

Rohrbough, Malcolm J. *The Land Office Business: The Settlement and Administration of American Public Lands, 1789-1837.* New York: Oxford University Press, 1968.

Rosenberg, Chaim M. *The Life and Times of Francis Cabot Lowell, 1775-1817.* Lanham MD: Lexington Books, 2011.

Rosenberry, Lois Kimball Mathews. *The Expansion of New England: The Spread of New England Settlement and Institutions to the Mississippi River, 1620-1865.* New York: Russell & Russell, 1962.

Roth, David M., and Freeman Meyer. *From Revolution to Constitution: Connecticut, 1763 to 1818.* Chester CT: Pequod Press, 1975.

Rothbard, Murray N. *The Panic of 1819: Reactions and Policies.* Auburn AL: Ludwig von Mises Institute, 1962.

Rothenberg, Winifred. *From Market-Places to a Market Economy: The Transformation of Rural Massachusetts, 1750-1850.* Chicago: University of Chicago Press, 1992.

_____. "The Emergence of Farm Labor Markets and the Transformation of the Rural Economy: Massachusetts, 1750-1855." *Journal of Economic History* 48, no. 3 (September 13, 1988): 537-66.

Ruschenberger, William S.W. *Summary History of the Academy of Natural Sciences.* Philadelphia: Collins, 1877.

Russell, Howard S. *A Long, Deep Furrow: Three Centuries of Farming in New England,* abr ed. Hanover NH: University Press of New England, 1982.

Seymour, Horatio. "Seymour, Horatio" Folder. 1816. Seymour Collection. Litchfield Historical Society, Litchfield CT.

Shalhope, Robert E. *A Tale of New England: The Diaries of. Hiram Harwood, Vermont Farmer, 1810-1837.* Baltimore: Johns Hopkins University Press, 2003.

Shelley, Mary. *Frankenstein, or the Modern Prometheus.* London: Oxford University Press, 1969.

_____. *The Letters of Mary Wollstonecraft Shelley.* Vol. 1. *"A Part of the Effect."*

Baltimore: Johns Hopkins University Press, 1980.

_____. *Mary Shelley's Journal*. Norman: University of Oklahoma Press, 1947.

Shelley, Percy Bysshe. *The Complete Poetical Works of Percy Bysshe Shelley*. London, New York: Oxford University Press, 1956.

Sherburne, Andrew. *Memoirs of Andrew Sherburne: A Pensioner of the Navy of the Revolution*. 2nd ed. Providence: H. H. Brown, 1831.

Six, James. *The Construction and Use of a Thermometer, for Shewing the Extremes of Temperature in the Atmosphere, During the Observer's Absence*. London: Maidstone, 1794.

Skeen, C. Edward. *1816: America Rising*. Lexington: University Press of Kentucky, 2003.

_____. "'The Year Without a Summer': A Historical View." *Journal of the Early Republic* 1, no. 1 (Spring 1981): 51-67.

Skinner, Walter R. "The Effects of Major Volcanic Eruptions on Canadian Surface Temperatures." In *The Year Without a Summer? World Climate in 1816*, edited by C.R. Harington, 78-92. Ottawa: Canadian Museum of Nature, 1992.

Smith, Andrew. *Gothic Literature*. Edinburgh: Edinburgh University Press, 2013.

Smith, Homer W. *Man and His Gods*. Boston: Little, Brown, 1952.

Smith, J.E.A. *The History of Pittsfield (Berkshire County), Massachusetts: From the Year 1800 to the Year 1876*. Springfield: C.W. Bryan, 1876.

Smith, Joseph. *The Pearl of Great Choice: Being*. Salt Lake City: Latter-Day Saints, 1878.

Spark, Muriel. *Mary Shelley*. London: Constable, 1988.

Steinberg, Theodore. *Down to Earth: Nature's Role in American History*. Oxford: Oxford University Press, 2002.

Stenza, Lisa. "Puritans: God's Ire Caused Cold Winters." *Hartford Courant*, 26 March 1984.

Steward, Joseph. *A Sermon Delivered at the First Presbyterian Church in Hartford, on the State Thanksgiving, November 28, 1816*. Hartford: George Goodwin & Sons, 1817.

Stilwell, Lewis D. "Migration from Vermont (1776-1860)." *Proceedings of the Vermont Historical Society* 5, no. 2 (June 1937): 63-246.

Stoll, Steven. *Larding the Lean Earth: Soil and Society in Nineteenth-Century America*. New York: Hill and Wang, 2002.

Stommel, Henry, and Elizabeth Stommel. *Volcano Weather: The Story of 1816, the Year Without a Summer*. Newport RI: Seven Seas Press, 1983.

Stone, William L. *Ups and Downs in the Life of a Distressed Gentleman.* New York: Leavitt, Lord, 1836.

Storr, Richard J. *The Beginnings of Graduate Education in the United States.* Chicago: University of Chicago Press, 1953.

Swift, Samuel. *History of the Town of Middlebury, In the County of Addison, Vermont.* Rutland: Charles F. Tuttle, 1971.

Tallmadge, Benjamin. Benjamin Tallmadge Collection. Litchfield Historical Society, Litchfield CT.

Tarbox, Increase Niles. *Thomas Robbins D.D.: A Biographical Sketch.* Cambridge: John Wilson, 1884.

Taylor, Raymond G. "The Importance of the Agricultural Revolution." *The History Teacher's Magazine* 8, no. 10 (December 1917): 342-44.

Thomas, David. *Travels through the Western Country in the Summer of 1816.* Auburn NY: David Rumsey, 1819.

Thomas, Isaiah. *Diary of Isaiah Thomas, 1805-1818.* Edited by Benjamin Thomas Hill. Worcester MA: The Society, 1909.

Thomas, Robert B. *The Farmer's Almanack, Calculated on a New and Improved Plan, for the Year of Our Lord 1816.* Boston: West & Richardson, 1815.

Thompson, Lawrance. *Melville's Quarrel with God.* Princeton: Princeton University Press, 1952.

Thompson, Noel W. *The People's Choice: The Popular Political Economy of Exploitation and Crisis, 1816-34.* Cambridge: Cambridge University Press, 1984.

Thompson, Zadock. *History of Vermont, Natural, Civil and Statistical. Vol. 2. Civil History of Vermont,* Burlington: Chauncey Goodrich, 1842.

_____.*History of Vermont: From Its Earliest Settlement to the Close of the Year 1832.* Burlington: Edward Smith, 1833.

Ticknor, George. *Life, Letters, and Journals of George Ticknor. Vol. 1.* Boston: James R. Osgood, 1876.

Tilton, Elizabeth M. "Lightning-Rods and the Earthquake of 1755." *New England Quarterly* 13, no. 1 (March 1940): 85-97.

Todd, Janet. *Death and the Maidens: Fanny Wollstonecraft and the Shelley Circle.* Berkeley: Counterpoint, 2007.

Trigo, Ricardo M., et al. "Iberia in 1816, the Year Without a Summer." *International Journal of Climatology* 29 (2009): 99-115.

Tuckerman, Joseph. Joseph Tuckerman Papers, 1799-1863, Ms. N-1682, Box 2. 15 July, 1816. Massachusetts Historical Society, Boston.

Tudor, William. *Letters on the Eastern States.* New York: Kirk and Mercein, 1820.

"Two Boston Puritans on God, Earthquakes, Electricity, and Fait: 1755-1756." *National Humanities Center Resource Toolbox: Becoming American: The British Atlantic Colonies, 1690-1763.* Web. 24 May 2011.

U.S. Department of Agriculture, Bureau of Animal Industry, 52nd Congress, 2nd Session. *Special Report on the History and Present Condition of the Sheep Industry of the United States,* Vol. 3124. Misc. Doc. No.105, 1892.

Udias, Agustin and Alfonso Lopez Arroyo. "The Lisbon Earthquake of 1755 in Spanish Contemporary Authors." In *The 1755 Lisbon Earthquake: Revisited,* edited by Luis A. Mendez-Victor, Carlos Sousa Oliveira, Joao Azevedo, and Antonio Ribeiro, 7-24. New York: Springer, 2009.

Van Doren, Carl. *Benjamin Franklin.* New York NY: Viking, 1938.

Vigilante, Stephen L. "Eighteen-Hundred-and-Froze-to-Death." In *Mischief in the Mountains: Strange Tales of Vermont and Vermonters,* edited by Jr. Walter R. Hard and Janet C. Greene, 97-101. Montpelier: Vermont Life Magazine, 1971.

Voltaire. *Candide and Other Writings.* Edited by Haskell M. Block. New York: Modern Library, 1956.

Walker, Bryce S. "The Earthquake of 1775: Science v. Religion." *Encyclopedia.com.* January 1997. Web. 15 March 2011.

Ware, Caroline F. *The Early New England Cotton Manufacture.* New York: Russell & Russell, 1966.

Ware, Henry. *Two Discourses Containing the History of the Old North and New Brick Churches, United as the Second Church in Boston.* Boston: James W. Burditt, 1821.

Ware, Henry, Jr. *Memoirs of the Rev. Noah Worcester, D.D.* Boston: James Munroe, 1844.

_____. "The Personality of the Deity." 23 September, 1838. *American Unitarian Conference.* Web. 7 July 2011.

_____. *A Poem Pronounced at Cambridge, February 23, 1815, at the Celebration of Peace Between the United States and Great Britain.* Cambridge MA: Hilliard & Metcalf, 1815.

_____. *Works of Henry Ware, Jr., D.D.,* Vols. 1-4. Boston: James Munroe, 1846, 1847.

Wayland, Francis. "Francis Wayland's Report to the Brown Corporation, 1850." In *American Higher Education: A Documentary History,* Vol. 2, edited by Richard Hofstadter and Wilson Smith, 478-87. Chicago: University of Chicago Press, 1850.

Webster, Daniel. *The Great Speeches and Orations of Daniel Webster.* Boston: Little, Brown, 1879.

Weiss, Thomas. "U.S. Labor Force Estimates and Economic Growth, 1800-1860." In *American Economic Growth and Standards of Living before the Civil War*, edited by Robert E. Gallman and John Joseph Wallis. Chicago: University of Chicago Press, 1992.

Welling, James C. *Connecticut Federalism, or Aristocratic Politics in a Social Democracy*. New York: New York Historical Society, 1890.

White, Alain C. *The History of the Town of Litchfield, Connecticut, 1720-1920*. Litchfield: Litchfield Historical Society, 1920.

White, Andrew Dickson. *A History of the Warfare of Science with Theology in Christendom*. Vol. 1. New York: Appleton, 1922.

Whiton, John M. *Sketches of the History of New-Hampshire, from Its Settlement in 1623, to 1833*. Concord: Marsh, Capen & Lyon, 1834.

Williams Family. Rev. Comfort Williams Papers, 1806-1840. RG 69:8. History and Genealogy, Connecticut State Library, Hartford

Williams, Mark. *A Tempest in a Small Town: The Myth and Reality of Country Life — Granby, Connecticut, 1680-1940*. Granby: Salmon Brook Historical Society, 1996.

Williamson, William D. *History of the State of Maine*, Vol. 2. Hallowell: Glazier, Masters & Co., 1832.

Wilson, C. "Workshop on World Climate in 1816: A Summary and Discussion of Results." In *The Year Without a Summer? World Climate in 1816*, edited by C.R. Harington, 523-56. Ottawa: Canadian Museum of Nature, 1992.

Winkless, Nels, and Iben Browning. *Climate and the Affairs of Men*. London: Scientific Book Club, 1975.

Winslow, Hubbard. *Rejoice with Trembling: A Discourse Delivered in Bowdoin Street Church, Boston, on the Day of Annual Thanksgiving, November 30, 1837*. 2nd ed. Boston: Perkins & Marvin, 1838.

Winterer, Caroline. *The Culture of Classicism: Ancient Greece and Rome in American Intellectual Life, 1780-1910*. Baltimore: Johns Hopkins University Press, 2002.

Winthrop, John. *A Lecture on Earthquakes, Read in the Chapel of Harvard College, in Cambridge, N.E., November 26th, 1755, on Occasion of the Great Earthquake Which Shook New England the Week Before*. Boston: Edes & Gill, 1755.

Wood, Gillen D'Arcy. *Tambora: The Eruption That Changed the World*. Princeton and Oxford: Princeton University Press, 2014.

Wood, Gordon. *Empire of Liberty: A History of the Early Republic, 1795-1815*. New York: Oxford University Press, 2009.

Woods, Leonard. *Address on the Life and Character of Parker Cleaveland, LL.D.* 2nd ed. Brunswick ME: Joseph Griffin, 1860.

Wordsworth, William. *The Complete Poetical Works of Wordsworth.* Cambridge MA: Riverside Press, 1932.

_____. *The Excursion: A Poem.* London: Longman, 1814.

_____. *The Prelude, or Growth of a Poet's Mind.* London: Dent, 1904.

Wright, John Stillman. *Letters from the West; or a Caution to Emigrants.* March of America Facsimile Series, Vol. 64. Ann Arbor: University Microfilms, 1819.

Young, Alexander. *Discourses on the Life and Character of John Thornton Kirkland and Nathaniel Bowditch.* Boston: Charles C. Little and James Brown, 1840.

Zeichner, Oscar. *Connecticut's Years of Controversy, 1750-1776.* Chapel Hill: University of North Carolina Press, 1949.

Zeilinga de Boer, Jelle, and Donald Theodore Sanders. *Volcanoes in Human History: The Far-Reaching Effects of Major Eruptions.* Princeton: Princeton University Press, 2002.

Zevin, Robert Brooke. *The Growth of Manufacturing in Early Nineteenth Century New England.* New York: Arno Press, 1975.

Zhang, Pei-Yuan, Wei-Chyung Wang, and Sultan Hameed. "Evidence for Anomalous Cold Weather in China, 1815–1817." In *The Year Without a Summer? World Climate in 1816,* edited by C.R. Harington, 436- 47. Ottawa: Canadian Museum of Nature, 1992.

Zielinski, Gregory A., and Barry D. Keim. *New England Weather, New England Climate.* Hanover: University of New Hampshire, 2003.

INDEX

Printed in the United States
By Bookmasters